T0369973

More Praise for

THE LAGOON

"At once ecological, scientific, and personal, James Michael Dorsey's book is a must-read for anyone interested in whales, whether gray or otherwise. Even those folks who have no interest in cetaceans might become converts after reading this excellent book."

—LAWRENCE MILLMAN, author of *Fungipedia*, *Last Places*, and *A Kayak Full of Ghosts*

"With his book *The Lagoon*: *Encounters with the Whales of San Ignacio*, James Michael Dorsey has written a beautiful testament effectively capturing and transmitting the aura and essence of these ancient and elegant creatures. I was mesmerized reading of his many encounters."

—WAYNE WHITE, author of *Cold: Three Winters at the South Pole*

"Dorsey's excellent book is much more than just the story of the magnificent gray whales. Important issues are addressed, from the need for conservation, to the many man-made threats these marine mammals face, including climate change, plastic pollution, captivity, and whaling.

"*The Lagoon* is a remarkable book by one of the best travel writers and naturalists, who has vast firsthand knowledge of the area and these intriguing whales. A vital read for anyone passionate about gray whales that also serves as a delightful guide to this stunning area, which people flock to each year to experience this unique phenomenon."

—LIZ SANDEMAN, co-founder, Marine Connection

THE LAGOON

ALSO BY JAMES MICHAEL DORSEY

Baboons for Lunch: And Other Sordid Adventures

Vanishing Tales from Ancient Trails

THE LAGOON

Encounters with the Whales of San Ignacio

JAMES MICHAEL DORSEY

DIVERSION
BOOKS

For more information, email info@diversionbooks.com.

Diversion Books
A division of Diversion Publishing Corp.
www.diversionbooks.com

First Diversion Books Edition: May 2023
Hardcover ISBN: 9781635768428
eBook ISBN: 9781635768947

Maps by Tim Kissel

Printed in the United States of America

10 9 8 7 6 5 4 3 2 1

≈

I wrote this book to take the reader on a mystical journey through land and time to the ancient domain of the gray whale. It is based mostly on two decades of notes, photos, and memories, so if I have erred in time or name, it was not intentional.

It is dedicated to Maldo Fischer and Johnny Friday, and especially to my friends and family at Campo Cortez, because it is their story, and I am honored they have allowed me to tell it.

Contents

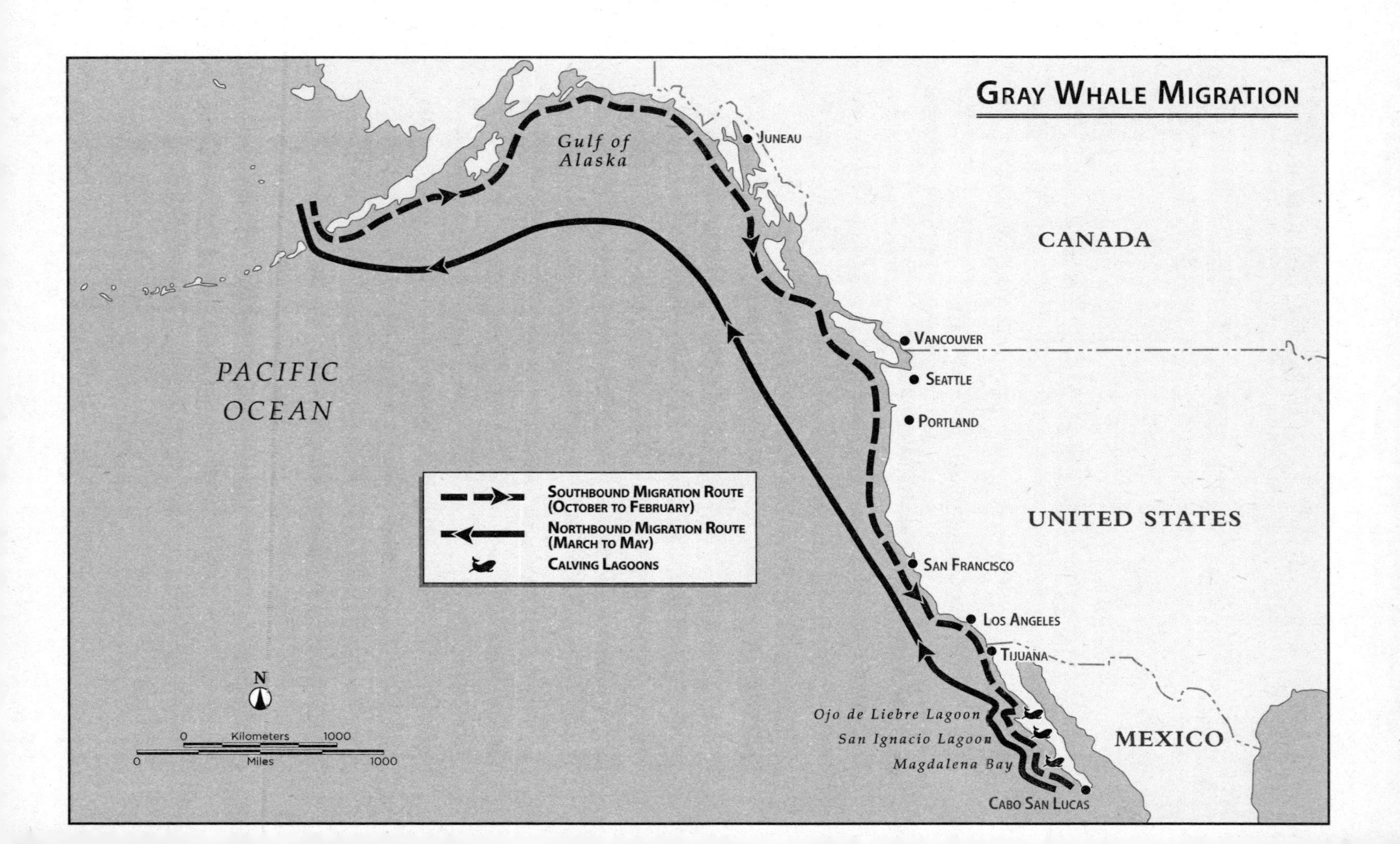
Gray Whale Migration
Juneau
Gulf of Alaska
Canada
Vancouver
Seattle
Portland
Pacific Ocean
Southbound Migration Route (October to February)
Northbound Migration Route (March to May)
Calving Lagoons
United States
San Francisco
Los Angeles
Tijuana
Ojo de Liebre Lagoon
San Ignacio Lagoon
Magdalena Bay
Cabo San Lucas
Mexico
N
0 Kilometers 1000
0 Miles 1000

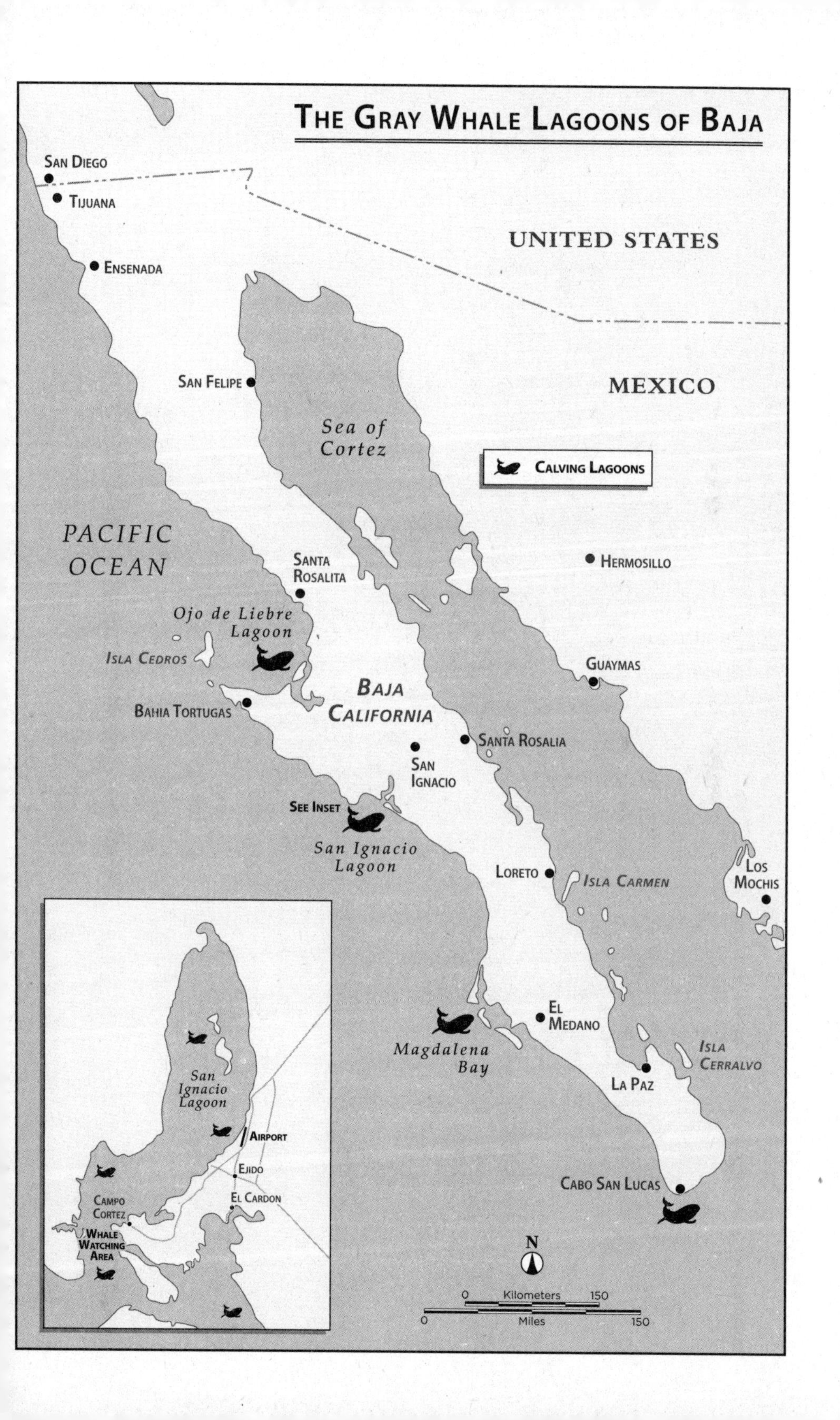

The Gray Whale Lagoons of Baja
San Diego
Tijuana
Ensenada
United States
San Felipe
Mexico
Sea of Cortez
Calving Lagoons
Pacific Ocean
Santa Rosalita
Hermosillo
Ojo de Liebre Lagoon
Isla Cedros
Guaymas
Baja California
Bahia Tortugas
Santa Rosalia
San Ignacio
See Inset
San Ignacio Lagoon
Loreto
Isla Carmen
Los Mochis
El Medano
Magdalena Bay
Isla Cerralvo
La Paz
Cabo San Lucas
San Ignacio Lagoon
Airport
Ejido
El Cardon
Campo Cortez
Whale Watching Area
N
0 Kilometers 150
0 Miles 150

THE LAGOON

ONE

Breaches

An immense gray back parted the water, rising like an island being born, while the mist from the whale's blow coated us like dew. She logged on top, then turned toward our panga, taking our measure. My fists were clenched in anticipation. This was a massive wild animal, an adult Pacific gray whale, a "devilfish," in its natural domain. Our only defense against such a giant was blind faith. I thought of the old stories, and it came to me that one of my own ancestors might have plunged a harpoon into one of hers.

I looked back at Maldo, hand on the tiller like the old sea dog he was. He was smiling, whispering to the whale in a voice meant only for her, using his magic to bring her to us. This was not fear, but love. The great back began to move in our direction, leaving a spreading wake on the surface.

A century and a half before our arrival, the waters of San Ignacio Lagoon in Baja, Mexico, ran red. The lagoon was a killing ground ruled by those who sought to gain riches from whale oil. Today, along with Scammon's Lagoon, it is part of the Vizcaino Biosphere Reserve UNESCO World Heritage Site, and the largest undeveloped wildlife sanctuary in the Americas. Along with Magdalena Bay, these three great nursery lagoons of central Baja are the only places on Earth where wild animals in their natural habitat routinely seek human contact.

San Ignacio's existence has never been without threat. In the lagoon, the old world collides with the new and is balanced on a fragile edge. In the past, the danger was only men with harpoons, while today it is men in business suits, a changing environment, people who love whales too much, and even other whales.

The whale sounded, sliding gradually beneath the surface where her gray and white camouflage could be tracked. Her flukes passed below us, revealing a missing piece the size of a white shark bite. Her tail was twice as wide as I am tall, but its movement was barely perceptible. Then the waters parted, and her massive head was next to us as she slowly rolled from side to side, giving us a thorough look. She was as long as two of our boats and outweighed us by close to forty tons. Her white, barnacle-encrusted rostrum betrayed her age as perhaps close to seventy, a fairly long life span for a creature so pursued by predators. She bore a lifetime of scars, most from shark attacks, and half a dozen prop strikes that she would not have felt because of her eight-inch armor of blubber. She hugged the boat like an old friend, diving below to rub her back on our keel as she rocked us back and forth, and when she came up, I looked deep into her blowhole, feeling her inner heat.

Living history was within my reach, an ancient and intelligent creature, unchanged in millenia, whose social order was beyond my understanding. Her ancestors roamed both land and sea for countless centuries before my own tribe arrived on the Earth. What stories she could tell. I found myself talking to her and realized I had never done so with an animal other than a dog.

She rolled on her side, and I caught my reflection in her iris, black, within a deep brown eye the size of my fist, and luminescent, with eyelashes any model would kill for. She did not blink, but held my gaze as I imagined her thoughts, willing me to understand. It was not the vacant look of a cow or pig that required no presence, but an intelligent look inquiring about the strange creature in her domain. I pressed my hand against her flank, finding it smooth and pliable like firm rubber while she returned the pressure, pushing against me. She came to each of us in turn, her mouth barely open, revealing aging, yellowed baleen, then submerged beneath. As she turned upside down below us, the panga began to turn in a slow circle. She was balancing us on her stomach, holding us in place with her pectoral flippers, and Maldo laughed as she played with us like a giant bathtub toy. When she had had enough, she disappeared as easily as her blow in a breeze, a thermal upon the water. How long she stayed I cannot recall, but it was not long enough. As Maldo started the engine, Irene's and my tears of joy merged with the sea spray.

I looked back once to see her spy hop (raise her head out of the water and rotate to look around), her massive head watching as we disappeared, wondering if she was calling us back or simply displaying her majesty as a queen of the lagoon. Motoring back to camp, I barely noticed the osprey passing overhead with a halibut in its talons, nor did I pay attention to the sea lion porpoising off our port side. Nature was alive and active all around us, but this one creature had captured me. I was lost in the moment, with no way of knowing it would be repeated countless times over the coming years. This was not the devilfish whose stories I had grown up with. One of the largest creatures on Earth had just gone nose to nose with one of the most fragile, and both of us came away with lighter souls.

≈

In February 1972, twenty-four years before my first contact with a gray whale, a Mexican fisherman named Jose Francisco "Pachico" Mayoral was fishing for grouper with his partner, Santos Perez, in a twenty-foot panga in San Ignacio Lagoon. A panga—driven by a pangero—is like a Boston Whaler. It is an open, fiberglass, round-bottom boat, the workhorse of Baja and a very seaworthy craft. In Baja the panga is the water transport du jour. The fishermen of the lagoon were used to the whales, but kept their distance, because they still feared them as lethal "devilfish."

Pachico loved to tell his story, and I had the pleasure of hearing it directly from him. A large female approached his boat and began to rub on its keel. Pachico and Santos, fearing the worst, hunkered down on the floor of the panga. When the whale would not leave, Pachico cautiously peered over the side of the boat to find himself looking directly into the eye of an adult gray, jet black with a brown iris and lashes as long as an elephant's. Gathering his courage, he reached over the side and touched the creature. The skin was slick and smooth. When the whale pressed against his hand, he jumped, startled by the animal's reaction to his presence. Pachico talked Santos into also touching the whale, but his hand was shaking so badly from fear that he never made contact. They both claimed the whale was heavily scarred from prop strikes, and in future seasons, this allowed them to recognize the same whale many times. That day, the whale stayed with them for a half hour. Try to imagine that feeling—an animal you have feared your entire life is suddenly caressing your hand.

That night, Pachico related the story to a skeptical wife and friends who either thought him mad from the sun, but over the next few days, other fishermen ventured out and began to touch the devilfish, and all reported the same result. This was world-shaking news for people who had shared the water for decades with an animal they feared. Had they always been friendly? Did we only need to reach out and touch them? Soon, wives, children, and friends were petting the devilfish. As they came to recognize individual whales, they began to assign them names. Something ethereal was taking form.

Reader's Digest published Pachico's story, and, in this version, he was curled up on the floor of the panga, making the sign of the cross and asking the blessed Virgin to spare his life. Later, Pachico would be featured in an IMAX film and mentioned honorably in Dick Russell's definitive book, *Eye of the Whale*. I met Pachico years ago in front of his house, the porch overgrown with plants and the house itself a construction of whatever flotsam the lagoon had given up. He wore a faded flannel shirt and a sun-bleached baseball hat, and I felt myself in the presence of a rock star. He could have been the poster boy for Hemingway's *The Old Man and the Sea*, with his weathered, leathery face a road map of tough times. Everyone who went to the lagoon wanted to meet Pachico, and he wore his celebrity with dignity. On a small sand dune across from the house, the aging panga from which he made first contact still sat, a victim of a half century of relentless sun. He spoke of the whale as part of his family. Many who knew him believed he possessed a special gift that drew the whales to him. One fisherman told me Pachico was a *brujo*, which normally would translate to "male witch," but in Pachico's case meant something closer to "whale whisperer." Once Pachico told me he often wondered if that first whale was an aberration or if perhaps it told other whales it was safe to approach people; I doubt that either case was true. Just like dogs who return with wagging tails to their masters who kick them, I believe gray whales, once hunted to the brink of extinction, were always friendly; humans just never gave them the chance to show it.

That first touch was soon almost as famous as God reaching out to Adam on the Sistine Chapel ceiling as its effect rippled throughout the world. Posters and postcards proliferated, illustrating "the touch," as it came to be known, and parallels were drawn between it and the famous baseball "Catch" by Willie Mays. Pachico passed away a few years back, but today, members of his family still own and operate Pachico's whale-watching tours in the lagoon, proud to share their father's story with clients.

The same year as "the touch," the Mexican government created a reserve and refuge area for migratory birds and wildlife in San Ignacio Lagoon, the United Nations called for a ban on worldwide commercial whaling, the US Congress instituted the Marine Mammal Protection Act, and the following year, the US Congress passed the Endangered Species Act, which included the gray whale. And it all started with an interspecies touch.

Pachico's legacy was thrust upon him. Almost overnight he went from a humble fisherman to a symbol recognized the world over as the godfather of whale watching. Today, all pangeros in the lagoon are graduates of a rigorous, government-run naturalist training course so they can take people on the water to meet and learn about the descendants of Pachico's whale.

Pachico never told me if he thought that whale had chosen him or if he thought it was a random occurrence. If the whale had approached a different panga, would the world still fear them today as devilfish? Why did that one particular whale rub against his boat? Was it cognizant enough to know that its actions could forever change how humankind viewed its species? I choose to believe that whale was, in its own way, something of an ambassador, sent to tell both humans and whales that the killing times were over, and it was time to start anew. Sometimes, it seems the tiniest occurrence, even unrecognized at the moment, often defines our futures.

After two-plus decades, I think of Pachico every time I touch a whale and have to believe he is smiling.

~

In the mid-nineteenth century, a whaling captain named Charles Scammon (1825–1911) was sailing out of San Francisco, California, but not as a long-range whaler like those out of New England who were gone for years on end in pursuit of humpback and deep-water sperm whales. He concentrated on hunting the shore-hugging gray whale that swam directly past his home base on their migration. His logs show that in 1855 he was taking whales in Magdalena Bay, Baja, while in 1857 he was whaling in Ojo de Liebre, a lagoon that would later bear his name. He first entered San Ignacio Lagoon in late 1859 after his whale hunting brother-in-law, Jared Poole, found it and told him it contained impossible numbers of whales. According to Poole, the lagoon was like a giant fishpond. Accounts written at the time said you could walk across the lagoon on the backs of whales.

It did not take long for the slaughter to begin.

Whale calves did not have enough blubber to make them commercially viable as prey, but the whalers would harpoon them first, knowing the cries of distress would bring the mother close enough for a kill. Yet, nothing prepared them for the wrath of angry forty-ton mothers. A gray whale's only defense is its tail, and a mother whale will wield it like a hammer and defend her threatened calf with the fury of a harpy. A large gray whale can knock an orca senseless with its tail flukes, and it was more than able to splinter the old-time, rickety whaling dories that were killing the newborns. For years, the lagoon ran red with blood as tens of thousands of whales were slaughtered, and no one knows for sure how many whalers died by the flukes of an enraged mother during that time. When the clashes between humans and beasts were exaggerated to sell newspapers, nothing was mentioned about mothers defending their innocent newborns. The gray whale simply became known around the world as a devilfish, a name that lasted until a half century ago. What a human trait to demonize an animal simply for defending its young from slaughter.

Ironically, in his later years, Scammon had an epiphany, and to recant for his sins, he not only retired from whaling, but became a leading advocate for the conservation of whales. In 1874 he wrote the book *Marine Mammals of the Northwest Coast of North America*. It was a financial failure, but over time it became a definitive non-fiction classic about whales that was only eclipsed in 2001 when environmental writer Dick Russell issued *Eye of the Whale*.

Today, not entirely accurately, San Ignacio's neighboring lagoon of Ojo de Liebre is more commonly known as "Scammon's Lagoon" for the man who once stained its waters red with blood.

~

Archaeological evidence tells us the whales Pachico and I had touched, and the whales that Scammon slaughtered, were descendants of those who shared this land and sea with the indigenous people, the Cochimí, who resided in central Baja, near San Ignacio, as early as ten thousand years ago. They are one of eight distinct tribal groups that occupied prehistoric Baja, and they were spread across the largest territory. The Cochimí were first encountered by Spanish explorers in the seventeenth century, and they initially hid from the strange-looking invaders. At that time, it is estimated that there were some sixty thousand indigenous peoples living throughout Baja. By 1762 that number had fallen to twenty thousand, and by 1800, less than six thousand indigenous peoples were still alive, thanks to the European introductions of measles, smallpox, and typhus, the eternal story of first white contact.

The Cochimí were hunter-gatherers who settled the coast to hunt and fish, but for religion and/or ceremony, they retreated to the closest mountains, the San Franciscos, where they painted and carved their history on monolithic walls, some of their renderings being carbon dated to more than nine thousand years old. There are almost four hundred painted cave sites in Baja, all believed to be painted by the Cochimí. Their creations carry great significance today. They left behind images of sea creatures in five separate caves, including fish—one halibut in particular is rendered quite accurately. There are turtles and bat rays, painted beautifully in local ochre. There is even one painting of what appears to be a very pregnant sea lion. Most important, there are multiple images of whales. Judging from the images' shape, lack of a dorsal fin, and small pectoral flippers, they are depictions of gray whales. One whale floats in space above other creatures of the land, while another is painted facing up, as if spy hopping on its tail. Yet another whale painted in red ochre is superimposed on a second in black, while a third hangs in the air with its pectoral fins drooping. The largest rendering appears to be a breaching whale. Whatever their creators intended them to be, they remain the first physical proof of the connection of the indigenous peoples of San Ignacio to whales.

So, what do these paintings mean? Other caves in the area depict traditional hunting scenes with people using bows and arrows, some showing antelope and deer being impaled by arrows, and others with animistic people raising their hands in supplication, and those are what one would normally expect to find in such remote areas. There is a cave with twenty-foot-tall people that has famously become known as the "Family of Man," but only the local caves had renderings of whales.

The closest cave containing a painting of a whale is twenty-five miles from the ocean. The terrain leading to this cave is brutal, filled with multiple species of thorny cactus, rattlesnakes, cougars, and wolves. The ascent offers only treacherous footing of loose shale and fragile sandstone. Such a journey for early people could be life-threatening. There had to be a deep meaning for them to have created these images so far away from their source. The Cochimí were not known to be seafarers other than close to shore in dugout canoes. They could easily have seen turtles, rays, and fish from shore or from a canoe, but what about the whales? Had they hunted and killed a whale? Had they encountered a dead whale washed up onshore? To depict them with such accuracy meant they had to have seen the entire creature. It is unlikely that they got into the water with such a giant animal, so this suggests they found dead whales onshore. It would have been the largest animal they had seen, so what did it mean to them? Was it a spirit being? Animists such as the Cochimí could easily have ascribed deital powers to such a creature. Perhaps transferring their images to the cave walls was their means of taking the power from the animal for themselves. The paintings' true purpose will never be known, but gray whales have held magical sway over the local people of central Baja for multiple centuries.

Despite their remote location, these painted caves are more popular with visitors than ever. Today they are overseen by Mexico's National Institute of Anthropology and History and are guarded and protected. In 1993, they were granted world heritage site status by the United Nations Educational, Scientific, and Cultural Organization (UNESCO). To some of us, they are altars and a link to the ancestors of those I have come to see. I choose to believe they were painted to keep them alive just as the descendants of the Cochimí protect the whales today.

Two

Bienvenido a San Ignacio

While San Ignacio revealed the gray whale to us, it was an orca in British Columbia that first led us to it.

We heard the whale blow before we saw it. The sound skimmed the water like a stone.

Its black dorsal sliced the surface, leaving a rippling *V* in its wake. We held hands to tighten our kayak raft—finding safety in numbers, and making us appear larger to a potential predator—hoping to get a look as it passed by. *It has to pass us*, I repeated to myself silently. Adrenaline pulsed through my heart like a drum, and I gripped my paddle with white knuckles. In a kayak, half of your body is underwater, so we were floating at eye level with the mightiest predator on Earth.

My wife, Irene, and I were paddling the Johnstone Strait in British Columbia, part of the Inside Passage, a kayaker's mecca and a summer freeway for spawning salmon and the resident orcas who prey on them. We never gave serious thought to actually "encountering" a whale; especially one with "killer" in its name, but one was coming toward us and fast! We watched its majestic black dorsal rise and fall, ever closer with each rising, its blows loudly powerful, like a steam engine coughing. It made one final dive, and then the world exploded! The great black head broke the surface in a mighty lunge not more than ten feet from our boat. There was a flash of pink tongue and huge white teeth; its six-foot dorsal towering above like a gleaming saber; a beast so large it seemed to blot out the sky. In that moment, I knew I was going to die.

We braced for collision as the whale sounded, rolling at the last moment so its dorsal cleared our keel by inches, and it glided silently by . . . directly below us . . . close enough to touch . . . and during that blink of eternity, two vastly different worlds merged. In that moment, dazed and unmoving, my mind raced with a thousand questions. It was the kind of moment I travel for, and on the rare occasions they present themselves, they end far too quickly. In the chaos of the moment, I had forgotten that it was our twenty-fifth wedding anniversary.

Now I have told two different personal stories, about two different whales, in two separate locations, so an explanation is in order.

~

That special wedding anniversary began this story, and it called for a special celebration. We decided to repeat our wedding vows in the venerable cathedral of Notre Dame in Paris. Those were the dark pre-internet days, so arrangements were made through letters and long-distance phone calls. In the back-and-forth deluge of red tape and finances required to fuel such an event, I lost my cool, called a religious bishop an obscene name, and was fired as a potential wedding client by the vicar of Paris. How many people can say they were fired from their own second wedding?

The big day marched closer as we searched for a special plan B, when an unsolicited ad appeared in our mailbox. It was a glossy photo of a man in a kayak, serenely gliding through a pod of orcas. It beckoned us to visit British Columbia and bask in nature's glory from a kayak, the ubiquitous modern transport of the Northwest Coast. The ad was so Photoshopped, even the whales were smiling, but it got my attention. A quarter century of marriage celebrated on the ocean in a kayak would be unlike anything we had done, and for that, it held an odd appeal. The fact that we did not know how to kayak was not an issue, and our complete ignorance of whales mattered even less. It would be an adventure, a learn-as-you-go trip, a true exploration. All I had to do was sell the idea to Irene. She has always been incredibly open-minded, and in fact, has traveled the world with me, often in less-than-comfortable conditions. But entering the ocean wild in a tiny, self-propelled, mostly plastic vessel not much larger than some fish was not high on her anniversary celebration list. Neither was camping in a northwestern rain forest, especially after losing Paris. More important, I had to gently suggest that flesh-eating, nine-ton whales would also be a possibility.

Needless to say, my celebratory choice won out. We would trade Parisian sheets for sleeping bags in the Canadian forest. Twenty-six years ago, we were set on a path that would have us kayaking and sailing for the next three decades from the southern tip of Baja to Alaskan waters and multiple points along the eastern rim of fire, always to find and learn about whales. It is a path that meandered often, but always led us back to Baja and the lagoon.

~

The name *California* has its own fascinating story. The one most commonly accepted comes from a sixteenth-century adventure novel titled *Las sergas de Esplandián* by Garci Rodríguez de

Montalvo. It tells the story of an all-black women's army that tolerated no males other than for procreation and then eliminated them. It states that they were led by a queen named Califia, and they were from an island known as Califor. The novel spoke of gold, possibly prompting later Spanish exploration looking for the fictional city of "Cibola," which legend said was built entirely of gold. Whatever the origin, Baja California means "lower California." Modern Baja has two states, Baja Alta and Baja Sur, upper and lower. We are going to travel through Alta into Sur. It was during these early explorations that the first written references to whales on the Baja coast came to us in the journals of Spanish explorer Hernán Cortés. They are the earliest known references to whales blowing since the Icelandic sagas (heroic prose narratives written around 1200).

The first known European to set foot in Baja was a Basque sailor by the name of Fortún Ximénez Bertandoña, who, under the command of Hernán Cortés, came in search of the gold of legend. Bertandoña led a revolt against his commander, and he and his mutineers landed near modern-day La Paz at the southern tip of Baja, thinking they had found the island Califor. Bertandoña was killed by local natives, and his crew returned to New Spain with a fictional story about black pearls but no gold. It was the lust for gold that lured numerous succeeding explorers to set sail for "Santa Cruz," the name given to the land by Cortés. From 1535 through 1539, Spanish expeditions in Baja encountered hostile native peoples, especially the Pericú, whose home was the southern end of the peninsula, and while the Spanish continued to explore the coast, they decided the interior was not worth the effort and relegated it to their missionaries to bring the native peoples into line with their rule.

Baja is a peninsula that resembles an elongated dogleg. It is 770 miles long (1250 km), 25 miles wide (40 km) at its narrowest, and

143 miles wide (230 km) at its widest. This long, narrow finger of rock is home to twenty-two volcanoes, making it one of the most seismically active areas on Earth. The land as we know it today was born some six million years ago when nature got angry enough to toss the Pacific and North American tectonic plates about, ripping a gouge in the Earth like massive scissors opening and allowing the ocean to rush in. It divided today's modern Mexico in two, and Baja resides on the Pacific Plate today where it continues to move west at the rate of about two inches per year. As the land separated, the peninsula's inner pressure rose upward to form a long spinal column of rugged mountain ranges, much like the Himalayas but on a smaller scale. From above, Baja is a long, bony finger that chinks and curves like the spine of a dragon, dividing the Pacific Ocean on the west from the Sea of Cortez to the east, a feral moonscaped land blessed with abundant wildlife.

The northwest shore of Baja has become one continuous development from just north of Rosarito, halfway to Ensenada. It is an anthill of condominiums, trailer parks, and mini resorts filled with retired American expats living the good life with an exchange rate of twenty pesos to the dollar. There is a complete lack of whale-watching companies in Baja Alta, which suggests most of the people may not even be aware of the massive wild animal migration that passes their front porches twice a year. There are no whale-watching tours until you reach Guerrero Negro several hours south. But this scene is just US spillover and not really Baja.

There is only one main, deteriorating road from north to south, linking the entire Baja Peninsula, and that is Federal Highway 1. Built in 1973, it is 1,060 miles (1,705 km) of twisting, winding blacktop with lanes so narrow, it is possible to be decapitated by just looking out the car window. The highway was never designed for today's modern cars, especially the extra-wide recreational vehicles so popular with Good Sam members who seem to be at every campground. Every few miles, motorists pass makeshift memorials just off the blacktop—a photo, flowers, and a prayer for the unfortunate soul claimed by his or her country's worst major highway. Flash floods and aging bridges are common causes for delays, not to mention cattle on the road. Once you hit the peninsula's elbow near Guerrero Negro, the turnoff to each of the central gray whale nursery lagoons is announced by an intact and complete skeleton of a gray whale. Southward, from there on, you are in the domain of the gray whale.

Northern Baja is green and touristy with dude ranches where you can have a bullfighting lesson with a young calf carrying tennis balls on its horn tips, then visit a local brewery. But as you head south, the modern world quickly recedes. The mountains begin to resemble an apocalyptic movie set, jagged and forbidding. Vast brown deserts of mesas, dry washes, and salt flats give it a sense of the primeval. Turkey vultures dot the sky, common as flies, soaring up steep-sided mesas on thermals over the scattered handfuls of pronghorn antelope and wolves that still roam free below them. Forests of giant cardon cactus grow forty feet tall and live for centuries, leaving behind a hard, woody skeleton widely used for construction. Kestrels and horned owls hollow out cactus to make comfy homes, but most desert inhabitants are nocturnal, coming to life only with the sinking sun. It is a land that captures one's soul at first sight, an infinite wasteland, seemingly devoid of life, yet teeming with it if you know where and how to look. The land screams "come to me" to those in need of nature. It is wilderness on an epic scale.

The three great lagoons host wetlands for myriad wildlife on land, sea, and the sky, their channels lined with sloughs and mangroves. The northernmost lagoon is Ojo de Liebre, which also contains Scammon's Lagoon. In the center is San Ignacio Lagoon, and in the south, Magdalena Bay. All three of these great lagoons have been sanctuaries and nurseries for gray whales since before recorded history.

I began working in San Ignacio Lagoon as a naturalist in 1998. I would bring groups of whale-watching clients with me, sometimes three or four trips a season. We would meet at a hotel in San Diego and take a private van to the border, where it was my job to secure tourist cards and get everyone over the border safely. Between bathroom breaks, snack stops, and photo ops, it was like herding cats, but my reward at the end was time on the water with the whales. To this end I learned to speak Spanish, but quickly found out that once I was south of the border, most people seemed to prefer English, not unusual since it is the international language of airlines.

But getting people over the border was not as simple as it sounds. In those days, there were lots of people with badges who had their hands out, so I always carried spare cash to make sure we did not sit for hours arguing with a self-important official in a suit. Clearing immigration and customs was simple once we got past the entry point. They had aging mechanical stoplights, the kind used back in the 1940s with alternating arms, one reading "stop" and another that said "go." You pushed a button, and if you got the green light, it was "Bienvenido a Mexico," but if you got the red light, you had to empty every item in your luggage and sometimes submit to a frisk search, not to mention a going-over of your vehicle with sniffer dogs and a bomb squad. At least in those days, we did not have to take off our shoes. Once inside Tijuana, we entered the familiar land of diesel fumes and the choking gridlock of traffic. The road south followed the old border wall long before its current expansion. By day, Tijuana is as busy as the Tokyo Ginza and just as peopled. Its shopping streets make you feel that if you were to faint, you would be carried along by the masses without falling. The van would then take us past the old bullfight stadium, mercifully no longer in use, then Caesar's Restaurant (where it is said the Caesar salad was first created), and finally to the public bus terminal to begin the overnight trip south. For twenty years, I made that round trip approximately thirty times. Those bus trips were an immediate immersion in the culture that set the mood for days ahead.

While I was taking paying clients on these annual trips, for me they were personal journeys to see friends, both human and animal. They were both pilgrimage and ritual, teaching me that journey and destination are the same. It is a difficult and time-consuming way of travel that is not for the faint of heart. A Mexican bus has been likened to a torture chamber on wheels, but whatever hardships must be endured along the way, no effort is too much to reach the whales.

Time slows in the Tijuana bus terminal, an aging, cavernous building and a time portal to old Mexico. Even in a metropolis as large as Tijuana, there is a mañana atmosphere compared to the corporate pace of cities to the north. The terminal's concrete and glass blockhouse is the size of an airplane hangar—a monument to 1950s Mexican architecture, a reminder of how slowly things change there. Inside, the smell of tortillas and mole mingles with the aroma of ammonia on linoleum floors. Sombreros, Stetsons, and cowboy boots are the dress code of the day for men, while elderly ladies wrap themselves in long, lace mantillas. The place always triggers a sense of belonging in me that does not translate easily into words, a feeling finely honed and nuanced over many years, somewhere between coming home and simple tranquility.

On a typical trip I pay my respects to the Virgin of Guadalupe, the patron mother of Mexico, whose statue stands tearful guard next to the entrance to the public restroom. I drop a two-peso coin into the pay slot that lets me revolve the steel turnstile and open the door marked "Caballeros" over the grinning stencil of a mustachioed man giving me a thumbs-up. Inside, I am pleasantly surprised to find flush toilets complete with paper, knowing they will probably be the last of their kind until I reach my destination.

In the main hall, I walk past the *cambio* money exchange that I have never found open, and I make my way to the counter with the oversized red, white, and blue letters spelling ABC, one of two national bus lines, and our chariot to southern Baja. Next, I wait while the young girl behind the counter writes tickets for my group by hand on a yellow legal pad with a dull pencil as she loudly snaps her gum.

Tijuana is an open city with no taxes, and it is here that the *braceros* and *agriculturos* of the south come to stock up on the trappings of modern society available from the large discount stores that have sprouted. The government-run ABC Bus Line is their shopping convoy home, and the waiting hall is full of people lugging big-screen TVs and assorted appliances on tiny folding luggage rollers. One ancient grandmother has three full shopping bags on each arm that cause her to roll like a camel as she walks. I watch a mother wrap her children in blankets on the cold steel chairs and try to make out the announcement on the PA, but it is mostly garbled static from a system half a century old. With my fellow passengers, I walk through the metal detector that beeps loudly at each of us but fails to gain the attention of the bored security guard, so the folding knife I forgot to take out of my pocket will stay with me through the night.

I typically took the 6:00 p.m. bus from Tijuana that was "on time" if it arrived any time between five and seven. The journey usually put us in San Ignacio by 9:00 a.m. the following morning for an average of fifteen hours a trip.

Outside, standing in line to board, the tiny grandmother clutching a canvas bag in front of me is startled when I greet her in Spanish. Once in her seat, she pats the one next to her for me to join her. She instinctively clutches my hand as the bus lurches from its stall. Her dress tells me she is from an older world, and clearly afraid of the journey ahead in this gigantic, modern machine. Perhaps she is from the desert, in the big city to visit a son or daughter, and now she must return. I assure her in Spanish that all will be well, and she nervously compliments me on my pronunciation.

Baja is not like mainland Mexico. It is older and set in its ways, preferring the old world to the new. The mainland is tourists and resorts, while Baja is travelers, explorers, and those open to the metaphysical. On the world scale, it is a tiny peninsula, but its deserts hold secrets and offer rewards. The jagged mountains appear shaped by an angry God, while tucked into its most remote corners are a people whose mode of living has changed little in centuries. They are people who give and share, and whose lives are filled with a mystical outlook often lacking in large cities. They are the same people whose ancestors turned back armored Spanish conquistadors with bows and arrows, the same people who spread Catholicism throughout the Californias, and the same people who today stand watch over waters that once ran red with whale blood. When you leave Federal Highway 1, horses and burros become the main mode of transport, and it is not unusual to find a homemade car made of cannibalized parts tucked behind an adobe.

Wooden doors of clay houses remain unlocked, because there is no crime among neighbors, especially since the neighbor is probably ten miles away. Cattle wander the highways with faces full of prickly cholla cactus, and cougars and wolves still lurk across a vast lunar landscape. Baja is a separate reality from my own life (and part of its allure is to realize that an imaginary line on a map is all it takes to divide such diverse cultures). That becomes obvious when our route takes us past the high steel wall that frames the border. It is covered with graffiti and seems eerily like one that used to divide Berlin. While crossing this wall is not as dangerous, it makes me yearn for a borderless world.

With the city behind us, the abuela (grandmother) releases my hand and with a timid smile of apology falls asleep with her head on my shoulder. She is going home now and is happy. The highway turns inland toward Ensenada, leaving the virgin coast behind.

Winding our way into higher elevations, the view is spectacular as the rugged coast twists back and forth before it disappears in the fog. We look down the dramatic sheer cliffs on numerous fish farms, revealed by the circular nets enclosing them. I recall on one trip watching a gray whale circling the nets, trying to figure out how to get to the giant smorgasbord just out of its reach. It must have been frustrating to think of cruising through such a gigantic bait ball only to find a nearly invisible net in the way. Gray whales do not usually eat while migrating, so that one may just have been curious, or as we shall see later, possibly, literally, starving. As the highway levels, the suburbs spring up followed by giant cranes that announce Ensenada as the largest port on the Pacific side of Baja.

The bus station in Ensenada is a hub for all buses traveling both north and south, and as a result, it is pure cacophony 24/7. The Ensenada bus station brings to mind the bar scene from *Star Wars* with a bit of *Mad Max*. The buses for which this is not the final destination will only be at the station for fifteen minutes. That is what the driver always says as people file out into the mass of humanity seeking late-night snacks or a *baño* (restroom). Most of the time, the buses will start up and begin slowly backing away after ten minutes, causing people to run or be left behind, sometimes pulling up zippers on the run.

Once back on the bus, it is a seventeen-mile drive to the second-largest marine geyser on Earth, which sits near the tip of the Punta Banda Peninsula. La Bufadora has several meanings in Spanish, the most appropriate being "snorter." It is a natural formation that resembles the blow of a whale on steroids. Huge local surf is forced into this partially submerged sea cave, where air and water pressure build until it "blows," sending a powerful blast several stories into the air. As always in Baja, at a place of natural history, there is a story to go with it, a legend, a myth, or often a metaphysical moment. The story of La Bufadora is that of a young gray whale separated from its mother on the southbound migration. The baby whale circled the area for days, crying for its mother, and when the rocks had heard enough, they captured the whale and made it their own. What you see today is the baby crying for its mother that never returns. (I was recently disappointed to find that since the last time I was there, this natural wonder has become a theme park with an outrageous entry fee.)

Another favorite story involves a fisherman who fell overboard while under the influence and was picked up by a gray whale, never to be returned. The legend says that on nights with a full moon, the fisherman can be seen as a silhouette, riding on the back of that whale. Both are sentenced to forever look for their way home.

Our journey continues through San Quintín, and I am reminded of a late-night phone call from a wreck-diving friend informing me of a stranded whale on the beach there. We drove through the night to meet my friend's bush pilot friend, who flew us over the beaches until we found it. It had not been dead long, but someone had already taken the entire head with a chain saw, probably for the baleen and skull that can bring serious money from aquariums or museums. Years ago, whales like this would have been left on the beach to eventually return to nature, but today, whale bones are considered valued collectibles.

In the silent cloak of night, Mexico has always carried a surreal air for me. In this Catholic land, organized religion has merged with superstition to create a belief system all its own, especially in the high mountains and remote deserts where I love to travel. This is the land of the *brujo*, witches, spirits, and demons, a land where people pray equally to Jesus in church and to syncretic images of Santeria in mud shacks. Even the grandmother sleeping on my shoulder wears a crucifix and a chicken's foot on the same chain around her neck because, like many people here, she occupies both worlds. I am hard pressed to think of anyone I know in Baja who does not have a metaphysical story or two. I have spent too much time there myself to dismiss their claims. I recall a midnight encounter in a coffee shop at Guerrero Negro when a stranger, arrayed head to toe in black, including sunglasses and hat, spoke cryptically to me about avoiding the moon, and then disappeared into a brightly lit and totally vacant street. Only then did I become aware that the moon was full and recalled so many legends of how the brujos traveled under its glow. More than any place I've traveled, Baja is an intersection of the physical and spiritual worlds.

On that note, in most countries, the biggest holidays tend to be celebrations of a country's independence, while Mexico celebrates death. "La Dia de los Muertos," or "Day of the Dead," is traditionally celebrated on November first and second. Families gather at cemeteries to create elaborate altars with photos of deceased relatives. They bring food, drink, and money to leave on the grave. Stores sell tiny skeletons that sing and dance and play musical instruments, and people will wear a mask or paint their face as a skull. It is an affirmation of the impermanence of life, and how quickly it passes. In Baja, more than the mainland, the skeletons sing and dance in celebration of how precious life is while serving as a reminder to savor every brief moment.

~

The first purple streaks of dawn slash the sky as we leave Highway 1 for the service road to Guerrero Negro, gateway to Ojo de Liebre and Scammon's Lagoon. Each telephone pole we pass hosts a wooden platform on top for the ospreys, fish-eating sea eagles that command the skies over all three lagoons, to nest without electrocution. On the beach side, we pass the second gray whale skeleton that announces this village as a major whale-watching destination.

Behind the bus station, the ever-present Virgin of Guadalupe, haloed by blinking Christmas lights, watches over the parking lot with outstretched hand. A faint whiff of marijuana comes to my nose, and stepping inside, two timeworn and gaunt vaqueros (cowboys) in straw hats are drinking coffee and passing a hand-rolled smoke. I buy a cold empanada so stale I toss it to a stray dog after one bite.

Because it is close to the ocean, Guerrero Negro is often bitterly cold at night. This evening I watch my breath rise in hazy clouds to disappear in the breeze. I think of the whales, only yards from me, but impossible to see in a blackness unmarred by city lights. If the wind should shift, I might hear them blowing, and I wonder if they can feel cold. I sense their closeness, but the moment to meet has not yet arrived. I think we will both be where we need to be soon. Above me, the Big Dipper sits low in the sky, the end of its handle pointing the way home for my return.

Guerrero Negro, "black warrior," takes its name from the American whaling ship *Black Warrior*, which operated out of Duxbury, Massachusetts. It was captained by Robert Brown, who managed to run her aground on a sandbar in 1858 in what was then called Frenchman's Lagoon. The current town was founded in 1957 by an American industrialist named Daniel Ludwig. Today, it is known for salt mines, which began with a small operation in 1954. It is the largest town in the Municipality of Mulegé in Baja Sur, with an approximate population of fifteen thousand. In 1957, Ludwig began a massive salt mining operation around the coastal lagoon of Ojo de Liebre, named Exportadora de Sal, S. A., of C.V., which translates to Salt Exporters, Inc. Over the years, it grew into the largest salt mine on Earth, with thousands of evaporation ponds spread over sixty-five square miles, exporting six million tons of salt annually to the United States, Japan, Canada, Korea, Taiwan, and New Zealand, a far-reaching product from such a tiny place. In 1973, Ludwig sold 51 percent of the operation to the Mexican government, and 49 percent to the Mitsubishi Corporation, the same one that makes cars and countless other stuff. Its location has long been debated as it resides within the protected Vizcaino Biosphere, and there are many who would like to see it gone. The salt works is massive enough to be clearly visible as you fly over it. Today, it employs more than a thousand workers, and, controversy aside, it brings in much needed cash for a very isolated people. The area around it, once a whale nursery, is now devoid of whales.

We leave Guerrero Negro to continue south as the morning light crawls over the horizon, spreading long shadows like spilled paint. No more than a mile from town, we pass an aging truck on the shoulder with a raised hood and smoke pouring from its radiator. A Green Angels truck is parked next to it. The Angels Verde are bilingual government employees who patrol the toll roads and major highways of Mexico day and night. They are similar to America's AAA (American Automobile Association), only on steroids. The Angels maintain a fleet of close to three hundred trucks that are in constant radio contact with several government agencies, and they provide mechanical assistance, first aid, basic supplies such as water, and will change a flat tire. As we drive past, I notice a smiling whale decal on the rear that says, "*Viajar sin peligro*," "travel safely."

An hour's ride south, the desert floor widens, and the road disappears into a cottony ground fog. The top of a distant volcano pokes through the haze, and massive cardon cactus slide in and out of sight, their upturned arms saluting as we roll past, a vast, silent army of escorts. Impressive as the massive cardons are, they are but one of more than a thousand cactus species indigenous to the Baja Peninsula. Suddenly, distant shadows become a herd of wild burros that cause us to brake hard enough to wake everyone. We laugh as the driver must exit and physically shoo them off the road. Kestrels are hunting insects in the morning haze, and the cactus appear to be stretching after the evening's sleep. Everyone crowds to one side, and cell phones are snapping burro photos. The quiet night is gone.

To the west, the San Pedro de Martir bleeds into the Santa Marthe and San Francisco ranges, which loom like an impassable barrier, their jagged peaks thrust upward like hands supplicating God. We are on the edge of the Vizcaino Biosphere, and if a dinosaur should appear, it would not be out of place. I look outside and wonder how anything can live in such gorgeous barrenness while knowing I am surrounded by life.

~

I believe it was 1998 when Irene and I first ventured into those mountains. We would call a rancho from town to say we were coming, and with how many people, and then take a van into the hills. Along the way, we passed shrines tucked into rocky alcoves, some with money and food, usually with a picture of Christ or a personal saint, all of them a testament to the religious fervor of the mountain people. We hired horses from the ranchers, but in the higher parts of the mountains we had to dismount and leave them with one vaquero because they were not sure-footed enough to negotiate the shale hillsides. The stirrups were encased with a steel boot to keep the prickly pear and jumping cholla cactus from shredding our legs. The horses wore leather bibs that covered their throat and chest against the fierce thorns of the cactus, as well as blinders to shield them from seeing the ever-present rattlesnakes. For the very isolated cave trips, we used mules instead of horses because, for every step up, the horses would slide down two. We were required to hire a vaquero guide who was invariably a stereotype from a western movie in a battered sombrero, leather chaps, a six-gun on his hip, and a Winchester lever action rifle in a scabbard. These guys are not dressed for the benefit of tourists, but are the real thing, born and raised in the mountains, used to cougars, wolves, and rattlesnakes, able to live off the land if necessary, and tough as leather. My favorite was named Martine, and he shaved each morning with his Bowie knife. They are cowboys because their fathers and grandfathers were cowboys, and they are fiercely proud of the title. The vaqueros always carried thick leather gloves and needle nose pliers for extracting cactus thorns. Their six-shooters were reserved for the snakes.

On an early trip when I only had two clients, we stopped for lunch at a rancho where I was told the elderly gentleman was a *brujo*, or a male witch. It is easy for anyone who is a bit different to carry that label. In this case, an elderly man with a sun-browned face resembling crinkled foil and shoulder-length white hair introduced himself and bid us enter. I recall his eyes being as black as midnight and his smile gentle. He shook my hand, holding it longer than necessary, and I felt something akin to a small electric shock. Thinking it to be static electricity, I dismissed it. Just then a mule acted up outside, rearing up and trying to kick the vaquero attempting to calm it. The old man excused himself, stepped outside, spread his arms, and appeared to whisper in the animal's ear. The mule instantly became docile and knocked its head to be petted. The old man returned to our lunch without a word, and I no longer had any doubts about him being a *brujo*. Baja holds more than its share of supernatural stories.

The first time I visited one of the painted caves, I sat with my vaquero guide taking in the landscape and laughing at jokes in my broken Spanish and his similar English. He was one of several teachers who showed me how to read an animal's sign by counting the number of toes in its track. He would pick up their scat, smell it, and tell me how long ago the animals passed by, and he would say it was a male or female by the depth of its print. Men like him are inseparable from the Earth, as much a part of the local landscape as the cougars and wolves. I mentioned the pictograph of a whale and asked him if he had ever seen a live whale. He was a modest man, born and bred in the mountains, who told me the biggest city he had ever visited had been Loreto, which at that time might still have qualified as a village. He told me it was because of the visitors he guided into the mountains that he decided for the first time to go down to the lagoon in hopes of finally seeing a whale and the famous Pachico Mayoral, who had a reputation as the first man to pet a devilfish, as he called it. He said that this fellow Pachico was a "*baleena susurraro*," or whale whisperer. He eventually found Pachico, but being a mountain man who had never been close to the ocean before, he balked at going on the water in a panga. The sea's vastness and depth shocked him. Unfortunately, the vaquero never gathered the courage to see a live whale.

Most of the mountain ranchos sit nestled at the foot of sheer sandstone cliffs. There were no trails, only cactus and steep shale rock walls. It is almost magical how the cholla will break free from the mother plant and "jump" onto any object that passes by as if they have a life of their own, embedding their barbs deep enough that they need to be cut out. We would often stop for a lunch of rice and beans, always offered before we asked, and were invited into homes that would have looked the same two hundred years ago—kerosene lights, natural rock fireplaces, animals' horns adorning the walls, and beautiful, handwoven, traditional rugs on the dirt floors. In one adobe I remember the stretched skin of a gray wolf covering a wall. Meals would always begin with the sign of the cross and a short prayer, regardless of one's personal beliefs. The mountain people were models of friendly hospitality and deep spirituality. Some of the larger villages held UNESCO status, and in each of these we always stopped to buy a large square of fresh goat cheese from a local *abuela* (grandmother). The first time we entered one of their homes, I saw a magazine photo of a gray whale taped to the wall behind a desk. She told me that during a particularly hard year, her son went down to the ocean to find work as a fisherman and, once there, he had seen the whales. He sent her a magazine article with the photo that she tacked to her wall. By now it should seem obvious that no matter the location, or situation, in central Baja, there is always a connection between the people and gray whales.

In these places, few people spoke English, and in those days my Spanish was minimal, but neither was necessary. People who live in such isolated areas share a universal communication of body language and gestures, and nothing works better than a smile. My body is now too rebellious to venture into the mountains in that mode of travel, but I am grateful for this opportunity to share them with the reader who may never have the gift of meeting them. The mountain people of central Baja are ancient and mystical, maybe as much as the whales themselves.

~

Back on the bus, the fog parts like a curtain and we pass low, flat mesas beyond which summits look like chocolate frosting on muffin tops. On the side of the road, isolated businesses in adobe homes display signs that read "*llanteria*" (tire store), "*comidas*" (food), and "*mechanico*" (no translation needed). Each sits in a perpetual cloud of highway dust and fumes, and all seem to own a dog that loves to chase cars and bark. It is isolated and lonely but there are always those drawn to such a life. It struck me that the means of earning a living in such places is usually dependent on those curious few who are just passing through.

In my twenty years of bus trips, there has always been a military checkpoint exactly ten miles north of San Ignacio. All vehicles are greeted by an oversized cardboard soldier holding a sign that says, "*Alto!*" Stop! The checkpoint is an impressive fortress designed to intimidate. Its main purpose is to search for drugs, and the Mexican army claims to have had much success over the years. There are sandbagged gun emplacements surrounded by coiled concertina wire and a jeep with a mounted .50-caliber machine gun. Soldiers stand about in the desert sun, their faces covered by buffs, and each toting an automatic rifle. Behind the barricades, a line of tiny pup tents attests to the hardship posting these soldiers have. The place gives the appearance of being ready for a major gun battle. Every vehicle must stop, while a serious-looking officer boards, checking papers as a drug-sniffing dog wanders up and down the aisle sticking its nose into carry-on bags. On rare occasions, they make everyone exit the bus and open all our luggage. I have never seen them harass anyone, nor ask for any money. The soldiers have always been polite, friendly, and professional, just doing their job, and any criminal would be intimidated if it were not for the snack bar operated by an overweight sergeant selling chips, cold drinks, and candy bars. On my last bus trip, they had even added whale-illustrated postcards for sale. On a lighter note, on two different occasions, the army's sniffer dogs refused to enter the bus because of the stench of both people and clothes after a week on the water with whales.

I have always appreciated the military checkpoints, which can pop up anywhere at any time in Baja. The army has always conducted themselves as gentlemen and especially show respect to elderly people, a quality I often find lacking in my home country. They have often asked if I am enjoying my time in Mexico. The Mexican army helps to keep the peninsula safe, and because of them, the reader may notice the lack of cartel activity that takes place on the mainland.

We round a hairpin turn, and in the tiny valley below us, looking like a sunken city, the adobe-tiled roofs of San Ignacio town make their appearance through the date palms. They are part of a tired and sun-worn village whose main industries are cement bricks and support services for whale watching. After having left the big city only twenty-four hours earlier, life in San Ignacio seems to be in slow motion. At the turn off of Highway 1, another skeleton of a gray whale welcomes you.

We have arrived at the gateway to a town that shows one face to the general public and another to those who know it intimately. In a three-hour drive, we will enter the lagoon of the same name where watches, iPhones, computers, and worries are all left behind. We pull into the dirt parking lot of the bus station, encircled by a small pack of barking dogs, and I spot Jorge leaning against his van, waiting for me even though we are three hours late. He has one cowboy boot on the bumper above a "Jesus loves you" decal, and his arm rests on the hood. He still wears the aviator shades I gave him two years ago. It is stifling hot under a van Gogh sun. A stray dog barks at a swirling dust devil, and I stare up at the familiar sign over the bus office. "Bienvenido a San Ignacio. You are exactly halfway between Tijuana and La Paz." No matter how many times I make the journey, each time feels as though it is the first. It allows me to slip into the mindset of the land and leave my regimented daily life behind. I smile at being back in Mexico. Jorge grabs my bag, and we are on the final stretch to see the whales.

THREE

Migration and Massacre

About 15 percent of gray whales are vagabonds. They spend their entire life in motion, even while sleeping. For them, movement is life. The Pacific gray whale spends the summer months feeding in the bountiful waters of the Chukchi and Bering Seas north of Alaska. In these open waters, massive opposing deep currents collide, bringing the whales life-giving krill and amphipods by countless tons. This is the midpoint of one of the longest migrations of any animal on Earth. Among whales, only humpbacks migrate farther. In the Chukchi and Bering Seas, the whales add to their blubber as fuel for the grueling seven-thousand-mile swim south they will commence in mid-December. The females will not eat during that swim. Indeed, while migrating, a nursing mother may lose a third of her body mass.

Why? Why does the gray whale leave its summer feeding grounds in these bountiful waters? Why do they leave their natural food source to practically starve for several weeks and run a gauntlet of several thousand miles past sharks and orcas and shore-based whaling stations to enter a tiny lagoon in southern Mexico? First of all, orcas are found the world over, so the whales are in no additional danger from predators while migrating than they are by staying in Alaska. Also, the arctic waters are too frigid for the young whales that have not yet developed their coating of blubber, and the Mexican waters are quite warm by comparison.

There are anywhere from eighteen- to twenty-two-thousand gray whales along the eastern rim of fire here in the Americas, a range that, while continually fluctuating, has held relatively steady for decades. Out of those, no more than perhaps three thousand will swim all the way to the Mexican lagoons; the rest will stop wherever they find a food source, or simply an area they feel secure, and many become residents of those areas, going no farther. That makes the comparatively small number of whales that enter San Ignacio Lagoon unique. The San Ignacio whales are but a fraction of those migrating, and an even smaller percentage of the total stock on the eastern rim of fire. The whales not born along the way return to the same lagoon they were born in, as surely as a baby goes to its mother's teat. Many calves will be born along the way with mother stopping only long enough to bear the calf, and then continue on immediately, pulling the calf with her in her slipstream as it is not yet strong enough to swim at mother's pace unaided, while others will wait for the warm, salt-rich lagoon to bring their offspring into the world. I have only seen one newborn within minutes of entering the world, and I only knew what was happening when there was a feeding frenzy of seabirds fighting over the afterbirth.

They go knowing they will be protected and safe. On a map, Ojo de Liebre Lagoon lies in the crook of Baja's elbow, which sticks out like a pointed dogleg and acts as a natural funnel for the southbound whales to enter. I have always wondered what inner voice tells a whale not to enter the inviting mouth of the first lagoon, but instead to swim all the way around the elbow of Baja to head farther south, but I am glad that so many do. Because of its location, Ojo de Liebre is typically more crowded with both animals and boats than San Ignacio, so I tend to avoid it. I also steer clear of Magdalena Bay to the south, as it is loosely regulated compared to the other lagoons, so the whales there are often harassed. The one time I did go to Mag Bay, as locals call it, I never got close to a whale. As soon as one was spotted, four or five overcrowded boats would charge it, and I found that intolerable. San Ignacio, on the other hand, receives only two hundred to four hundred whales per season, a number that the lagoon seems to support with ease. Their smaller numbers also make identification easier than in the more crowded lagoons.

We do not know for sure why the same whales return to the same lagoons every season, but it is most likely because they navigate by underwater geography, so they need familiar landmarks that over time become locked in their memory, allowing them to travel most of the way on autopilot. Being the slowest swimming whale (3–5 knots, 3.4 mph–5.7 mph), the gray must use any means available to survive. The littoral provides rock formations, kelp, and even giant fish schools that the gray whale uses for camouflage. If a predator is near, it will hang in the water, holding its blow, called *logging*, or descend into the maze of a kelp forest. Gray whales have been known to hide beneath the hull of sailboats at the approach of a predator. On more than one occasion, I have watched them stir up the silty bottom with their tails to hide in the murky cloud till danger has passed. They are masters of disguise, and their natural, uneven markings also serve to hide them. If you examine the coloration patterns of a gray whale, you will see noticeable similarities to camouflage used today by our military. It is one of the smaller contributions of whales to humans.

I have been on the water near the mouth of the lagoon when whales enter. It is not just a celebration of survival, but also a homecoming, a chance to see old friends again after many months of separation. They show their joy openly, often breaching and nuzzling one another, all the while aware that their journey is only half finished for the year.

So, what inner call do they answer? Toothed cetaceans such as sperm whales, belugas, and all dolphins travel by echolocation, meaning they use a tiny burst of air to send out an electronic signal that bounces off an object and comes back to them as a vibration, allowing them to form a mental picture of what is coming their way, but gray whales do not have this ability. Perhaps, much like the first wolves that approached humans on the way to domestication eons ago, whales may be attempting to enter the world of humankind in much the same way, but there is no way of knowing definitively how long they have migrated. When a species repeats an action over centuries, I believe it enters their DNA. They migrate because their ancestors did, just as the fishermen of the lagoon are fishermen because their ancestors were.

Originally, there were no natural predators in the lagoons, so the whales' babies were safe while still helpless and basically in whale school. The lagoons contain a much higher salt content than the open ocean, and so the calves are buoyant and have little chance of drowning. And yes, a whale can drown just like a human since we are both air-breathing mammals. Gray whales are born with little natural instinct, so from the moment mother pushes a baby to the surface to draw its first breath, it receives constant attention in a 24/7 school for survival. The lagoon has always been a sanctuary without predators, but recently, that situation has changed, and as I stated earlier, the lagoon has never been without threats. In 2022, it seemed that violence caught up with the whale nursery in the lagoon, and this will be addressed later, in chapter 10.

We may never know the actual reason for their migration; there may be many different ones, but I understand why they do it. I have the same inner voice that brings me back to the lagoon each season on my own migration. According to UNESCO, there are some seven to nine million hereditary nomads on Earth, those who travel as a way of life. It only makes sense that denizens of the ocean travel for the same inner reasons. Gray whales are nomads. It is simply an impulse deep inside that must be followed by both human and whale.

≈

Many people think of whales as living dinosaurs, but those species came and went some 230 million years ago. The first creature that might be called a whale did not appear for another forty million years. It was an artiodactyl, or hooved animal, closely related to cows and deer and even giraffes and camels. It walked on land and breathed through nostrils on the front of its face just as you and I do. Sometime around six million years ago, this creature decided it had had enough with walking around on land and began the glacially slow process of transitioning to a sea creature. We know this because today's whales still have a vestigial pelvic bone, and skeletons left behind by primitive cetaceans called archaeocytes show the progression of the nostrils from the front of the face to the top of the head, where today's whales breathe through a blowhole. About fifty million years ago, another animal, known as Pakicetus, joined the cetacean family. It looked nothing like a whale as we know it. From skull specimens, it was determined that the animal was about one to three feet (one meter) long, and six to seven feet (two meters) tall. It had four functioning legs and a long snout, could move its head independently of its body (only the Beluga whale is capable of this movement today), and had a full set of teeth. Just like modern human noses, Pakicetus's was at the tip of its snout. While originally a land animal, later generations became fully aquatic, and it was associated with cetaceans because of similarities of the inner ear to modern whales, but that is about all they have in common with today's whales. Pakicetus has been mostly forgotten as the skeletons left behind resemble a modern dog more than a prehistoric whale.

Another close relative who lived about fifty million years ago was Ambulocetus, which could live both in and out of the water. It thrived near estuaries and had large feet that resembled flippers, but it had a regular flipper-like tail that it used like a rudder for swimming. A mere forty million years ago, the Durodon appeared. It was about fifteen feet in length with actual flippers and tiny rear legs. It was totally aquatic and gave live birth underwater. It is the closest known relative to the modern whale. In the course of just over ten million years, cetaceans completely transitioned from land- to sea-dwelling creatures. That may seem like a long time, but in geological terms, it was a mere hiccup in eternity. The above-mentioned whales all had teeth, and as they evolved, many—such as the orca, sperm whale, dolphin, and porpoise—retained them. But about thirty-five million years ago, some whales' skulls flattened, and they developed a feeding filter called baleen that grew down from the roof of their mouth. It resembles the hairs of a whisk broom. Humpback, blue, and gray whales all have baleen that is keratin, the same as our human fingernails. All of this occurred before the indigenous Cochimí people of central Baja appeared in the fossil record about ten thousand years ago, so whales, in some form, may have been visiting the Mexican lagoons for some 190 million years before the first identifiable humans showed up.

Whaling, the deliberate killing of the largest animals on the planet using a harpoon, can only be traced back directly about five thousand years. In the late 1990s, two harpoons were found in what is known today as the Democratic Republic of Congo. They were carbon dated at up to ninety thousand years old, but they never killed a whale. It is believed they were used to spear giant catfish. They had a barbed point and a loop for attaching a rope and these Semliki harpoons, as they are known, are considered to be the earliest known examples of a modern harpoon. It is the rope that separates a harpoon from a common spear. Spears are thrown with the hope of killing upon impact, while a harpoon is thrown with the intent of tethering the animal to a boat to tire it and wear it out before killing it.

The first people to kill a whale are lost in the pages of history. Most likely they surrounded the whale with small dugout canoes, trying to force it toward shore where it could be killed with spears. They may have woven nets to entangle the animal. I have always wondered about the motivations for this first kill. Was it done out of fear, or an attempt to provide food? If it was the latter, it is hard to imagine primitive humans killing such a giant in order to eat when much smaller and easier game was accessible. Certainly, the Cochimí were not the only people on Earth to encounter whales. Early humans in more remote parts of the world were most likely making the same contact as the Cochimí, only with different local whales. They would have been subsistence whalers, taking only what they needed to survive at great personal risk, and there are still many such people in today's modern world. For those people, the taking of a whale's life is a covenant between humankind and animal in which the animal is usually thanked for giving up its life.

There was ceremony before and after the hunt. The nations that have a history of subsistence whaling have had negligible effect on the population of their indigenous animals.

It was the advent of steamships and hydraulic gear systems in the mid-nineteenth century that created industrial ships capable of slaughter on an unheard-of level that took many species to the brink of extinction. Whalers of that era had a different mindset, and there was nothing spiritual about it. It was large-scale, wild-animal slaughter for profit on a level beyond that of any human warfare in the history of the world. The only comparable display of senseless killing would be the annihilation of the American bison on the open plains. The bison roamed seas of grass while whales roamed seas of water. Both died in the millions for no reason other than profit.

We know that around two thousand years after the Cochimí painted the ochre whale on the Baja cave wall, the Basque people of northern Spain were taking North Atlantic right whales off the coast of France in the Bay of Biscay. That would have been early in the twelfth century. In 1605, a British sailor who was part of an expedition into what today is the state of Maine published a written account of local native people conducting a whale hunt from their dugouts. By 1610, England was taking whales in the waters of her American colonies, and the pilgrims of the *Mayflower* claimed to have witnessed whales "playing hard" directly off their bow. That same year, Japan was on the cusp of opening to the outer world for the first time, and by the late seventeenth century they had developed a national taste for blubber that exists to this day. From that time on, Japan, more than any other nation, has had an insatiable craving for whale meat. By the way, North Atlantic right whales got their name from their interior bladder that kept them afloat when killed, making them the "right whale" to hunt. All other whales sink upon death unless their mouth is immediately sewn shut to prevent water from entering. On the old whaling ships, it was the sail master whose task it was to sew up the mouths of deceased whales. It was also the sail master who sewed deceased sailors inside shrouds for burial at sea. The former was meant to float, and the latter to sink.

The first commercial whaling operations out of America sailed from Nantucket, Massachusetts, in 1712. The last American whaling ship under full sail was the *Wanderer* out of New Bedford, Massachusetts, and it was blown aground by a hurricane in Buzzards Bay in 1924. John R. Martin was the last of the old-school sailing whalers working out of New Bedford as late as 1927, although he was now on a modern industrial ship. The last shore-based whaling station did not cease operations until 1960. In the years in between, the sperm whale was the kill of choice. Sperms have an enormous head and a narrow lower jaw full of teeth, and they are known to battle with deepwater squid. Sperm whales carry lots of oil that was coveted for candles and lamp wicks, while the bones were used for everything from ladies' corsets to umbrella stays. Even more valuable was ambergris, a pliable, wax-like substance produced in the whales' digestive system that at first smells like fecal matter, but with aging takes on an earthy flavor much coveted by perfume makers even today. A large, solidified chunk of ambergris could and still can command thousands of dollars on the open market.

Most whaling ships were hunting on the other side of the world from their home ports, so whales had to be flensed (skinned) and their blubber rendered at sea. But what exactly is blubber? It is a thick layer of fat beneath the top skin of most marine mammals. It is thicker and contains more blood vessels than the fat of humans. Some marine biologists consider blubber to be a connective tissue uniting the skin of the whale to its internal organs. It is everywhere in their body except for the flippers, tail flukes, and fins. Blubber helps keep the animal buoyant because it is less dense than ocean water, and it insulates the warm-blooded whale from the cold. It is a thick, oily layer that stores fats and proteins. Blubber has long been part of the regular diet of arctic peoples. Thickly sliced hunks of skin and blubber were called muktuk, and it is a rich source of both vitamins C and A.

Once the whale has been flensed, the blubber is rendered, or slowly cooked, over a low flame in gigantic iron pots called trypots that bring to mind a witch's cauldron. As the blubber melts down, it turns into a waxlike substance that is called *whale oil*. That is the final ingredient so coveted that tens of thousands of whales died to light our lamps and put fragrance in our perfume.

Rendering whale blubber on old-time sailing ships was hard, dirty, stinking work. Once the whale was totally rendered, the oil was stored in wooden barrels, sometimes years at a time before the ship would return to its home port. The head, flukes, and flippers were discarded unless kept as gruesome souvenirs. Fortunately, in more modern times, petroleum and natural gas have made whale oil obsolete as a fuel source while vegetable oil has replaced whale oil in food products. Unfortunately, whale oil is still coveted today as a chemical lubricant.

On today's factory ships, the flensing is still mostly done by hand, but the rendering process has been speed up with high-tech machinery. Sometimes several whales will be hung by their tails, upside down on the side of the ship, all waiting their turn to be turned into oil. It is a disgusting sight and a vulgar display of humankind's relentless pursuit of money. The animals die to keep our machines running smoothly. There is no spiritual connection with the animal, no ceremony to thank it for giving up its life, no prayers to purify the event. It is simply greedy slaughter at its worst.

The massive teeth of sperm whales were, and still are, popular collectors' items, especially when scrimshawed by a sailor. The word *scrimshaw* can be traced back to the twelfth century and is believed to have been a surname from the French word for a "fencing master," *escremisseor*. The act of etching scrimshaw is believed to have begun in the late eighteenth century as a way for sailors to kill time on multiyear voyages. Teeth of their sperm whale victims were plentiful and large, by far the largest teeth of any whale, and the whaler/sailors etched lines in them with a needle or knife point and then filled them in with lampblack to create images as delicate as a dandelion's thistles. It was the early whalers who first gave the world both tattoos and scrimshaw. Whale teeth were in great demand by the public as collectibles and souvenirs of the dangerous journeys the collector was unable to make him- or herself. Some scrimshaw artists wielded such prowess that they gained wealth and fame in their own lifetime. According to the New Bedford Whaling Museum in Massachusetts, the first known American scrimshaw artist was Edward Burdett, most active around 1805, and the best known was Frederick Myrick, who was at the height of his powers around 1808. The work of both of these whale artists is on display in the museum today and considered some of the finest offerings of this traditional American art form. Fortunately for sperm whales, but not so for walruses, the Marine Mammal Protection Act outlawed the taking or possession of whale ivory, so the scrimshaw industry has turned to walrus teeth in their stead. Today, the possession of marine mammal parts is a maze of laws and contradictions. Much depends on what animal part in what state. Despite their significance, it is important to realize these works of art only exist because whales died.

As is often the case with a species in danger, their body parts became ever more valuable as public demand for them rose. Whalebone was in great demand, not just as carved curio pieces, but to make musical instruments, knife handles, chess sets, walking canes, you name it. If it came from a whale, it sold, thus ensuring their continued imminent demise. While whale teeth may no longer be legally traded or sold, the more intricately carved pieces of scrimshaw command tens of thousands of dollars in today's black markets. Gray whales were spared this indignity because they have no teeth, but there still exists a black market for trinkets, especially women's jewelry, made from aged baleen that resembles opaque plastic in older whales.

Faster swimming whales such as blue, sei, or fin whales were too swift, and so avoided being slaughtered. Humpback whales were also too agile and acrobatic to make for easy prey. All of those fast-swimming species could leave a ponderous whaling dory in their wake. It was the glacially slow gray whale—swimming an average of three to five knots (3–6 mph)—that suffered the brunt of the assault. In full panic mode, it can make seven knots at most (8 mph). Unlike the dolphin, it does not echolocate, so it must hug the coastline to navigate, making it an easy target. It did not carry the oil of a sperm whale, but its blubber could be rendered, and its bones made into objects to amuse people, putting it near the top of the kill list.

Before hydraulic-fired harpoons, a crew member had to stand in the bow of a rowed dory on a constantly pitching ocean and plunge a razor-sharp harpoon into the flank of the whale. That meant the crew had to be directly alongside the animal, moving at panic speed, while hoping the whale did not change course into them. Keeping one's balance was next to impossible, all the while trying to aim a fifteen-foot lance and hurl it with enough strength to penetrate almost a foot of blubber. If it strikes a rib, it will most likely not become embedded. The harpoon would be attached to hundreds of yards of strong rope that was tied to the boat. If the rope was not coiled properly before the hunt, it could fantail around the boat and often entangle crew and drag them into the water. The frantic whale would take off at full tilt, pulling the boat and all on board in what came to be known as a "Nantucket sleigh ride," and it was violent. The lucky whales were the ones that died instantly. The whale would run flat out until tiring from loss of blood and the weight it was pulling. Sometimes empty barrels would be attached to the rope, making it all but impossible for the whale to sound. Hundreds of men died in the act as the whales defended their young. In the late nineteenth century, whaling was the most dangerous job on the planet with the largest financial rewards if successful.

For two centuries, the only known and documented recorded attack by a whale on a ship took place on November 20, 1820. The next occurred in 2022 and will be discussed shortly. In an event that seemed impossible at the time, on that day in 1820, a sperm whale repeatedly and deliberately rammed a whaling ship until both it and the ship sank. Since then, there have been a few unverified accounts of whale assaults, but nothing like the ordeal of the whaleship *Essex*.

The *Essex* sailed under the command of George Pollard, a ship's captain from a venerable whaling family out of Nantucket, Massachusetts; her first mate was Owen Chase, a veteran whaler. Hunting off the coast of South America, her dories had harpooned one whale, and it was crying loudly when a larger than normal sperm whale was sighted. Crew estimates said the whale was eighty-five feet long while a normal sperm might reach sixty-five feet.

With the whaling ship's dories deployed, the large whale charged perpendicular to the *Essex*, and rammed it head-on. It did so repeatedly, until it split its own head open, and both the whale and ship sank together, into a churning crimson sea. This tale, written by Owen Chase in his book *The Wreck of the Whaleship* Essex contends that, "The whale was trying to protect its fellow whale that had been harpooned, and its cries of distress were clearly audible." I personally believe that Chase thought it was coming to the defense of its own. Whatever the reason, the story of the *Essex* is true and Owen Chase's book is still in print.

I have tried to put myself inside Owen Chase's mind as he bobbed on the water in the dory, watching his mother ship slide beneath the waves. Half of the *Essex* crew went down with the ship while the other half rode the swell on the open ocean, not knowing if the whale was dead. They must have expected another attack, as the hunters had become the hunted. Fortunately for them, the whale unleashed its fury against the *Essex* itself, and not its crew. If the whale had attacked the dories, then the world would never have known the fate of the *Essex*. But Chase's statement about the whale coming to the aid of a stricken companion stood out because it is one of the earliest written references to a whale feeling pain and empathy by one who would know. In those days, there was no consensus of whales being anything more than large, dumb fish, whose only purpose was to serve the needs of humankind. To me, the whale sacrificing itself in defense of a companion was the highest manifestation of empathy in a sentient being. What could be more humanlike? Chase's words were the earliest I could find expressing the possibility that a whale is a sentient being.

Chase and three fellow survivors spent almost seven months in their whaling dory and resorted to cannibalism to survive. When he returned to England on the rescue ship *Eagle*, his days were occupied with charges levied against him. He was also ridiculed because no one thought a whale capable of attacking a large ship, let alone sinking one. He was labeled a fraud and publicly humiliated. The ordeal followed Chase, and he suffered headaches and nightmares. He horded food in his attic and ended up in a mental institution for the final eight years of his life. I prefer to think that Chase died out of remorse for his actions, but then it was a different time with a different public mindset.

Coincidentally, a member of the *Essex* crew was a friend of the son of the author Herman Melville, who at that time had published only a couple adventure novels. That crew member gave Melville's son a copy of Owen Chase's manuscript, and Melville shaped the story into *Moby Dick*. Though a lackluster seller when published in England, and not selling well in the United States, today it is considered a great American novel, an allegory of good versus evil, the maniacal sea captain blindly pursuing a whale to kill it. I found great irony in *Moby Dick* because Captain Ahab believed the whale was a calculating killer. That would imply he believed the whale was intelligent. It is also interesting to note that Scammon's middle name was Melville.

People of the *Essex* era knew little of whales, only what they read in the papers, and most never saw a whale unless they lived near the water. There were no whale-watching tours, no scientific expeditions. Whales were nothing more than a food source, something to be killed and eaten and their bones repurposed as household tools and knickknacks, and if they fought back, as we've seen, they were labeled devilfish. With literature such as *Moby Dick*, it is easy to understand how an animal could be demonized in the eyes of the public. For good or bad, the whaleship *Essex* and *Moby Dick* made the entire world aware of the power and majesty of whales.

Only recently, reports have come from Spain and Portugal that orcas have been ramming sailboats. Are these attacks, or a response to high frequency radar and sonar, or some other hidden reason caused by modern technology? A story related to me by an old sailor friend offers my personal best guess about this unprecedented behavior. An orca breached the hull of a trimaran on its way from Hawaii to Tahiti. The sailor patched the hole and continued, but with the main hull full of water, one of the amas (side pontoons) was out of the water while sailing. From below, the silhouette of the boat hull could easily resemble a mother whale, and the side ama might look like her calf to a predator from below.

Even though these events happened in different oceans, orcas range hundreds of miles in a day and have numerous lingual dialects. It is possible that one orca rammed a boat, thinking it to be a whale and passed the idea on, as ramming anything is definitely a learned behavior. Just as is often the case in a shark attack, once the predator hits a surfboard and realizes it is not food, it usually spits it out. This could be the same thing happening with the orcas. That is only one possible reason for the recent rammings.

Another possibility came to me recently after watching a video shot by a researcher friend. It is known that orcas will hunt a seal hauled out on an ice floe by forming a line and swimming full speed at the floe, stopping at the final second to generate a wave to wash the seal off the floe into open water. In this video, the seal was on a floe much too large to be dislodged by a wave, so the orcas dove underneath and began to ram the floe from below with their heads until they broke up the floe, dumping the hapless seal into the water. No less than Sir David Attenborough has seen this video and believes no more than a hundred orcas currently practice this behavior. It would certainly help explain the ramming of boats in hopes of achieving food.

~

In 1737, the British South Sea Company introduced explosives to whale hunting, dispatching a fleet with cannon-fired harpoons. While these worked, no one thought about the whales filling with water and sinking before they could be harvested—and that is exactly what happened. Over the next century, several inventors "improved" the harpoon to include a barb that lay flat along the shaft while entering the animal, but being on a hinge, flared out after impact, locking the harpoon head inside the whale's body and allowing it to be tethered to the ship. Later in the 1820s, British inventor William Congreve, who previously introduced the first rockets to the British army, developed a rocket-propelled harpoon that impaled the whale before exploding. He believed the internal explosion would generate enough gas to prevent the whale from sinking. It did not.

Whaling shifted into high gear in 1868 when a Norwegian whaler named Svend Foyn invented a harpoon cannon that fired a razor-sharp shaft into a whale with an explosive charge embedded in the point that also exploded inside the whale, with the same results as those of Congreve. With two such devices of mayhem now deployed, people had refined the killing of innocent animals to an industrial level never before dreamed.

A modern whaling harpoon is usually launched from a deck-fired cannon. Its point contains a charge of penthrite, a compound similar to nitroglycerin designed to kill instantly, but that is not always the case. Once the barbed point is lodged inside the whale, the mother ship can simply reel in its prey, often with the whale still alive. Modern anti-whaling organizations such as Greenpeace and Sea Shepherd (its full name is Sea Shepherd Conservation Society) have often reported the agonizing cries of harpooned whales.

Since these animals are normally silent, it must take tremendous pain for them to physically cry out loud. As technology advanced, processing ships could harpoon a dozen whales a day, hang them on the sides of the ship like trophies, flense, and boil down their oil and blubber while still at sea. Technology made killing whales quick and easy on the open sea.

The mid-nineteenth century has often been referred to as the "golden age" of whaling. The Massachusetts whaling town of New Bedford was called America's richest city, if not the whole world's, while nationwide, roughly seventy thousand people were working in the industry in one form or another. At that time, America's population was only twenty-three million people, of which more than three million were slaves. Dangerous as it was, it was high-paying work for people of color in the nineteenth century, and some seven hundred black men entered the industry during those years, with sixty-three rising to the rank of captain. Among them was William Shorey, who was called the "Black Ahab" for his prowess. There was great irony in the fact that on land these black captains could not safely look a white man in the eye, but at sea, they could order one to be flogged. Beginning in the late eighteenth century, the main source of light in American homes was the waxlike liquid called spermaceti, contained inside the massive heads of sperm whales. It is believed to act as a receiving platform for their echolocations. The spermaceti organ can hold as much as 500 gallons (1,900 liters). There is also evidence that spermaceti helps control buoyancy. Science would indicate that being warm-blooded keeps the temperature of its spermaceti up, thinning its density, which, in turn, helps the whale to float. When it wishes to dive, it will ingest seawater, cooling and solidifying the spermaceti, which allows the sperm whale to dive deeper than any other whale. To rise, it expels the water, and its own inner heat raises the temperature of the spermaceti. On a side note, the whale got its name from old-world whalers who thought this liquid resembled human sperm.

It is fascinating that only toothed whales possess this ability. Sperm whales are also unique in their production of ambergris, that flammable waxlike substance produced in their digestive system that, as it ages, takes on a fragrant, earthy scent treasured by the perfume industry as a fixative to make scent last longer. It is produced in the whale's gastrointestinal tract, and since the beaks of giant squid—a sperm whale's preferred food—have often been found inside globs of ambergris, some scientists believe it is produced to help them pass hard or large objects through the tract. On rare occasions, whales will regurgitate ambergris and large chunks of it may wash ashore where it resembles dark boils from a tree. Dogs seem unusually sensitive to the smell and are often used to track it. In a letter to a friend, Benjamin Franklin stated that spermaceti "offered a clear white light, could be held in the hand without softening, and its drops do not make oil stains like common wax candles." He included in his letter that spermaceti candles "last longer and need little or no snuffing."

Nevertheless, the decline of the industry came quickly. The proliferation of other nations entering the whaling wars, especially Norway, the introduction of fossil fuel oil, and the American Civil War all conspired to defeat traditional whaling. Still, the new, modern "factory" ships accounted for 10 percent more dead whales in 1931 alone than in the entire nineteenth century.

At the same time as ocean, or pelagic, whaling, shore-based whaling along the western coast of America was in high gear, especially in Monterey, California, along the hereditary migration route of the gray whale. Most of those early shore-based whalers came from Portugal, initially to seek their fortunes in the recently found goldfields. When they became discouraged by the sheer numbers of those seeking the precious metal, many of them switched to shore-based whaling operations. They brought their whaling knowledge with them, as many had also worked on the long-range ships out of New England in earlier times.

≈

While numerous species of whales were taken from shore-whaling stations, it was the slow-swimming, shore-hugging, migratory gray whale that was the prime victim. The first shore-based station began in Monterey around 1850–1851. By the end of the century, there were an estimated fifteen to twenty stations along the coast from Trinidad in Northern California to Santo Tomas in upper Baja in the south. By 1912 the Arctic Oil Works was operating an enormous refinery near San Francisco, and photos show acres of gray whale baleen stacked like corn stalks, waiting to be turned into buggy whips, ladies' corsets, and umbrella ribs, to name only a few of its uses.

The exact number of these stations is in debate, because these operations often came and went in short time. These companies usually comprised a captain, first mate, a couple of boat steerers, a cooper (barrel maker), and roughly eleven men. Two boats, each manned by crews of six, would put in together. Living conditions at these stations were a hygienic nightmare. The whalers worked for "lays," which might be compared to today's stock options, with each man getting a cut of the overall profits. A gray whale could surrender up dozens of barrels of oil at a time when it was selling for about forty-five cents per gallon. A barrel held about thirty-five gallons, so the profits were enormous for their time.

A unique communication system of flags was employed between spotters onshore and the crews of the boats that had put in. The dories would hang out in the water waiting for an unwary whale to approach. If a whale was spotted, the shore-based men would lower a red flag. This was a signal for one of the boats to begin circling. When the boat was facing the direction of the approaching whale, they would wave the flag. If the boat crew was unsure of the whale's location, they would dip their dories' sail until they were redirected to the proper location. In order not to spook the approaching animal, they would belay the oars and sometimes paddle silently using only their hands. Once the whale was harpooned, it inevitably led to the aforementioned "Nantucket sleigh ride," and in more cases than not, the whale would fill with water and sink before it could be reached. No one knows how many whales were killed close to shore and lost due to sinking.

Charles Scammon was familiar with the shore-based stations and wrote the following: "At the point where the enormous carcass was stripped of its fat, arose the whaling station, where trypots were set in rude furnaces, formed of rocks and clay, and capacious vats were made of planks to receive the blubber."

I've witnessed how this rendering technique was applied myself, near my current home. Just south of Monterey, at Point Lobos, a shore-based whaling station displays its original, late-nineteenth-century harpoons, flensing knives, hooks, and other whaling tools, including a trypot. The grounds are filled with whale bones, and many are inlaid in the flooring, especially the flat compression discs from a gray whale's spine. Inside, the building is filled with historic photos, and in 1994 it officially opened as the Whaling Station Museum. It is open to the public today.

A similar wharf and processing plant built in 1866 lies fifteen miles north of Monterey. Moss, later renamed Moss Landing, was constructed by Texas millionaire Charles Moss, along with a Portuguese whaler named Cato Vierra. From 1919 to 1926, shore-based operations, controlled by California Sea Products, hunted whales from there, with the first whale being taken in January 1919 by a Norwegian captain named Fredrick Dedrick, commanding two steam-powered chase boats from the mother ship, *Hercules*. The meat from these early whales was used as chicken feed, and the bones were ground to make meal. As whale stocks diminished from overhunting, most of the shore-based stations were shut down by the early 1900s, but the last station of its kind, near San Francisco Bay, only ceased operations in 1971.

Having worked at these local stations, I have tried to imagine what it was like for migrating whales. We know they communicate with one another, so it was likely they knew what was awaiting them all along their route, and yet they continued their almost suicidal swim twice a year. Without echolocation, they were forced to hug the coast, making them easy targets. That speaks to the power of their inner need to migrate.

By the late 1930s, the whale census on the California coast counted only a few dozen gray whales migrating south, although some may have taken the long route on the other side of the Channel Islands. The census concluded the depleted population of the so-called Eastern stock along the rim of fire was due to over-whaling. At that time, it was estimated that the world's combined whaling fleets were killing up to fifty thousand whales per year. Within a decade, blue and sperm whales were on the edge of extinction.

The Atlantic stock had been counted extinct since the mid-nineteenth century, when Gloucester, Nantucket, and New Bedford, Massachusetts, were in their heydays as whaling towns. With only a few hundred Pacific grays left, what saved them from extinction was World War II. As the war spread across the globe, food rationing became the order of survival, and whale meat was a luxury by any standard. Industrial nations that had previously been vigorous whalers were forced to put operations on hold so they could concentrate on killing people rather than whales for the duration of the war. At that time, there were also no organized whale-watching tours. The general public had little interest in whales unless they lived near the coast, where seeing them was inevitable. No one was watching them with a scientific eye, and no one was counting them.

It was not until about 1948 that people who lived along the coast began to notice the increased numbers of migrating whales once more, and they found to their astonishment that hundreds were once again following America's western shoreline heading south. It took about six years of man killing man to give gray whales a respite. As America awoke from those dark years, people began to take stock of what had not been available for so long, and the fires of curiosity were lit once more, but they were not strong enough to cease commercial whaling for another two and a half decades. While people took interest in the newly returned migration, most people still thought of the whales as dumb animals with no real purpose. Slowly, the public began to view whales as more than food and commercial material. The famed local marine biologist Ed "Doc" Ricketts was not just John Steinbeck's best friend, but a Monterey superstar, one of the earliest marine biologists who provided aquariums and laboratories with specimens from around the world. His writings were introducing an isolated public to the wonder of sea creatures. Monterey was transitioning from a whaling town to a whale-watching town. With such a large and fertile bay, the Monterey Bay Aquarium Research Institute (MBARI) was founded in 1987 by David Packard of Hewlett-Packard fame and maintains its labs and offices at Moss Landing, California. Besides the world-class aquarium, they maintain three deepwater research vessels, two ROVs (remotely operated vehicles), and a devotion to the preservation and research of this almost thirteen-thousand-foot deep bay. Today, Moss Landing is home to three whale-watching companies, one from which I only recently retired, Sea Goddess. I took great comfort in educating people about the fact that in that tiny bay where whales were once slaughtered, they are now being watched and enjoyed as the Lord's fellow creatures.

~

Before World War II, there was no central highway through Baja. As recent as 1966, the paved road ended seventy-two miles south of Ensenada. From there it was unpaved, undeveloped washboard, filled with potholes and rocks, and only the most intrepid explorers ventured as far south as the lagoons. The war also expanded public aviation, and as the first clipper air ships began to cross the oceans, remote places like San Ignacio Lagoon were suddenly accessible to those who could afford it. Among those seeking to experience unexplored Baja was the famous movie actor Errol Flynn, who began to visit San Ignacio Lagoon in 1948. Flynn was a keen amateur marine biologist following in the footsteps of his father, who was a professor of marine biology at the University of Belfast in Northern Ireland. He was not just an action movie star but lived a lifestyle that paralleled his film roles. That year he arrived in the lagoon with a helicopter and film crew, hoping to produce a documentary about gray whales.

The crew flew low over the animals hoping to herd them into shallow water where they could be better filmed. As they approached, the whales, who had been docile to that point, began to churn the water with their tail flukes. A member of the film crew said: "The grays churned the water with flukes and fins until their wakes became swirling cauldrons of foam. Faced with such displays of angry power, the helicopter lifted to a safe altitude." This was the first known account of the whales reacting to an aircraft. The crew remained for several days but kept their distance until the final day of filming when the whales had had enough. On this day, a panga the crew had rented for filming was rammed by a gray so hard it threw the fishermen off their feet. The creature continued to ram it until they frantically retreated to water too shallow for the whale to continue its pursuit. Was that an attack or self-defense? At that time, whales were still being hunted and were particularly anxious at the approach of humans, let alone a giant noisy bird harassing them from the sky.

Baja was being discovered by tourists, explorers, and scientists, eager to get out and travel after years of war. Among them was one of the world's leading cardiologists, Dr. Paul Dudley White, personal physician to then president Dwight Eisenhower and a cofounder of the American Heart Association. Through the years, Dr. White had studied various animals in hopes of unlocking the secrets behind human heart disease. In 1956, Dr. White sailed to Scammon's Lagoon on a former Coast Guard ship that was now a research vessel named *Dorado*, provided by Douglas Aircraft, with the intent of making the first cardiograph of the heart of a gray whale. The lengthy operation, which carried over into a second year, was filmed as the documentary *Operation Heartbeat 1956–57 California Gray Whale Research Project 29494*. (In those days, gray whales were called "California" gray whales; today they are recognized as "Pacific" gray whales. Also, the more popular spelling today is *gray*, and not *grey*.)

The team launched a powerboat flying the flags of their sponsors, the Douglas Aircraft Company and the National Geographic

Society. They were armed with barbed harpoons and crossbows that fired darts, all of which trailed long wires back to a telemetering craft that in theory would give a printout similar to a modern EKG. If successful, it would be a black-and-white chart of a whale's heartbeat. While attempting to harpoon a whale, the creature turned on them, severely damaging their pursuit boat with its tail flukes and negating the collection of any data at that time.

In 1957 the team returned with a helicopter, apparently having learned nothing from Errol Flynn's earlier expedition from the sky. They now had official support from the US Army and Mexican government. This time, high-powered rifles were used to fire the darts and harpoons with wires attached to a trailing, automated powerboat that would follow the whale by remote control. It was the first time the whales of Scammon's Lagoon had a helicopter fly low over them. As the chopper passed overhead, the whales scattered in panic. It would seem at this point that Owen Chase from 120 years earlier had more empathy for the animals than Dr. White, who only saw them as a subject for study.

To get the desired cardiograph, two harpoons would have to penetrate the whale deeply. They both did. The harpoon tips were designed to disengage after a few minutes but there was no record that they did. Dr. White's team got the cardiac information it set out for. It was the first true scientific information about the gray whale, predating by five years data obtained from two captive whales named Gigi and J. J., who will be discussed later in chapter 9. There is no way to know definitively if the gray whales of Baja contributed to conquering human heart disease.

In the late 1960s, a film crew of the famous oceanographer Jacques Cousteau was in San Ignacio Lagoon to film grays. His crew had a reputation for being environmentally friendly, but in those years, the term was interpreted quite differently than it is now. The crew chased a female for more than an hour, sometimes getting right up on her tail flukes. The whale breached, destroying the boat. When I asked local fisherman about the validity of the story, one told me a crew member had died in the incident, while another told me the whale only gave them a warning. Apparently not many remember the incident except as a story, because at the time, no one in the lagoon knew who Jacques Cousteau was. The internet is surprisingly void of additional information. What would the world have thought if the headline read "Cousteau Divers Harass Whales"? Today, diving is not allowed in the lagoon except by scientific research permit. For one thing, there is poor visibility due to the constantly swimming whales stirring up the silt, and the whales have shown repeatedly that they easily spook when a diver enters the water.

Were these incidents actual attacks by the whales, or frightened animals attempting to defend themselves? I hope that none of them would take place today, but they illustrate how the public perception of these whales perpetuated the title of devilfish for so many years. It took the touch of a humble fisherman, Pachico Mayoral, in San Ignacio Lagoon to pass the devilfish name into history.

FOUR

A Blue that God Made

There are no signs announcing that you have entered one of the most pristine wilderness areas in all the Americas. El Vizcaino Biosphere Reserve is marked by an imaginary outline on a map showing the borders of an ancient and unique land that Mexico is determined to keep and preserve. It was singled out for protection in 1988 and is the largest environmentally protected area in all Latin America and the pride of Baja. At 143,600 square miles (2.5 million hectares), it sits squarely in the center of the peninsula, covering a quarter of the land, including two of the three main gray whale sanctuary lagoons. It was named for a Spanish soldier, explorer, entrepreneur, and diplomat, Sebastian Vizcaino (1548–1624), who in 1593 sailed from the mainland of Spanish Mexico, across the Sea of Cortez, with the mission of proving Baja California to be an island. In 1596, on behalf of the viceroy of New Spain, Gaspar de Zuniga, Vizcaino explored the area that now bears his name. In 1602, Vizcaino sailed up the west coast of modern California as far north as Mendocino, following in the wake of his predecessor, Juan Cabrillo, who preceded him by six decades. Vizcaino's journey proved the peninsula was not an island, but his main contribution to exploration was to declare much of the findings and place names of Hernán Cortés to be inaccurate and to rename them. Most of Vizcaino's place names remain to this day, especially this nature preserve that bears his name.

The biosphere has sixteen distinct areas that cover the mountain range of Tinajas de Murillo, the Vizcaino Desert, islands of Laguna Ojo de Liebre, the shore of Guerrero Negro Lagoon, and other small islands within the biosphere. It also claims an area 3.1 miles (5 km) offshore for the protection of the migrating whales. This offshore area contains sixteen small islands. Within its total boundary there are marine, coastal, desert, freshwater, agricultural, and mining mini-ecosystems, each one a unique kingdom of nature.

Federal protection covers every rock, shell, cactus, and tree branch within its boundaries, and to harass an animal of any type, especially a whale, is to awaken the wrath of the gods. Within the biosphere's borders are several seventeenth-century missions, hundreds of cave paintings dating back ten thousand years, more than a half-million migratory birds, forests of towering cardon cactus, and two sanctuary lagoons where gray whales have given birth and nursed their young for centuries. Also, at less than three people per square mile, it is the least populated region in all of Mexico. While there is no physical boundary defining the biosphere, the people who live within it carry a sense of responsibility and national pride. In 1993, the entire biosphere, including the sanctuary lagoons, and the cave paintings, were each granted UNESCO World Heritage Status.

≈

The turnoff from Highway 1 to the town of San Ignacio is perpetually guarded by the official icon of the town famous the world over for gray whales. The intact skeleton of a young gray whale is surrounded by a white picket fence giving it a homey, garden-like feeling, and at any given moment, day or night, there will most likely be people taking selfies next to it. Whale bones are ubiquitous on the western shore of Baja. They line trails and paths, they adorn gardens, decorate patios, and fill shelves in private homes. They sit outside restaurants as objets d'art, and are used as teaching aids in whale camps. They are part of the landscape. I have always found them to be beautiful, aesthetic objects, and admit that in my earlier years I brought many of them home from Baja with me before learning that they were protected. I can only take solace in the fact that I use the bones I have to educate people about the whales they came from.

Before we enter the town, I will relate a personal story that took place across the street from that whale skeleton. I ran into Dick Russell, the previously mentioned author whose seminal book, *Eye of the Whale*, has become the definitive work on whales. We run into each other every now and then, but meeting in the middle of Highway 1 in Baja was surreal. We stood like a scene from a movie, reminiscing, and watching a brilliant sunset. At the time I never dreamed that one day I would also author a book about whales. That was the kind of moment a traveler prays for, and I believe some whale karma was at work.

The road into San Ignacio meanders through a date palm oasis that during winter months produces the sweet fruit the town is known for. It crosses the beautiful Rio San Ignacio from which the town takes its name. The trees are heavily populated with blue and night herons, greater and snowy egrets, and the ubiquitous turkey vulture that many laughingly call the "state bird of Baja."

The shore is lined with cormorants who fly underwater to hunt fish with the same wing action they employ in the air. They must air dry their wings after a dive before they can "fly" again, so they open them wide and look as though they are waving as you pass by. When the wind is mild, kayakers drift lethargically below, occasionally landing a fish from its emerald, green water. From town, the river wanders south for dozens of miles, frequently disappearing underground to surface again farther on, its location betrayed by the oasis of palms that identify where it has surfaced. The town brings to mind a movie set, a sleepy, typical Mexican village built with a square around a central church.

The area was first scouted in 1706 by Jesuit Friar Francisco Maria Piccolo who is credited with having founded the town. The first mission in Baja was founded in 1697, and from then until 1767, seventeen different missions were constructed along with numerous "visitas," or sub-missions. For three centuries, Baja has marinated in Catholicism. It would seem that every town and village, regardless of size, hosts a Catholic church, and the religious fervor of the rural people is palpable throughout Baja. The local palms and generous citrus trees of San Ignacio were planted at the start of the eighteenth century by a Jesuit missionary named Friar Juan Bautista de Luyando who also constructed the original church in 1728, built on an ancient Cochimí settlement known as Kadakaamán, which in Cochimí means "Arroyo of the Reeds." The land had been occupied by the Cochimí for centuries before the first padre arrived. The church's actual name is Mission San Ignacio Kadakaamán, and it is the focal point of the town.

The original church, like everything else, is made of basalt rock, a constant reminder of the nearby active volcanoes. Thanks to nonstop seismic activity, the church has been rebuilt a number of times even though its walls are four feet thick. In 1767, the area was part of Spain, and the then ruling king, Carlos III, thought the Jesuits too intertwined with the Catholic inquisition and ordered them expelled on January 29 of that year. The king then decided that ten Dominicans would fill the vacancies left by the departing Jesuits. The current church was constructed in 1786 by a Dominican friar named Juan Gomez. The mission closed in 1840, but during much of the eighteenth century it was a meeting hub for Jesuit, Franciscan, and Dominican missionaries. If you look closely, on the right-side wall as you face the church, you can see numerous musket ball holes left from when Napoleon III's troops used the nave for target practice in 1862. Today, a daily Mass is said in the church, and it is not unusual to see pilgrims climbing its steps on their knees in supplication.

Inside the main nave there is a statue of a Peruvian-born saint named Martin de Porres, who seems to have no connection to any of the other religious figures but holds a unique place in the history of the town. Local people I have asked make him sound like a Mexican Saint Francis. De Porres was looked down on because of his mixed birth, but he was a colorful character credited with the ability to talk to animals. One story attributed to him tells of the monastery where he was stationed that was overrun with mice. When the friars decided to lay traps for them, the legend goes that de Porres spoke to the mice and told them not to return lest they be killed. After that, no more mice were to be found. Many local townspeople believe that de Porres's ability to communicate with the animals extended to the whales.

The church grounds include a beautiful garden of bougainvillea and indigenous cactus that resides directly next door to a human-carved, subterranean cave that houses one of the town's least-known attractions. It is the wonderful but minuscule National Institute of History and Anthropology Museum. Inside, next to numerous artifacts of early life in Kadakaamán, there is a life-size reproduction of the giant mural pictograph known as the Family of Man, mentioned in chapter 1. It also contains facsimiles of some of the whale pictographs. Besides learning the local history, you can arrange for a tour on mule or horseback into the local mountains to see the real thing.

Irene and I used to take clients to visit the Family of Man cave at the end of our whale trips, especially to show clients the pictographs of the whales and their connection to the area as a pleasant way to end their journey through Baja. On one of these trips, Irene was backing up to take a group photo when she lost her footing and went tumbling over the cliff side. Fortunately, she did not fall very far but she landed on her knee, shattering the cap. It so happened that one client was an acupuncturist from Israel who had her needles with her. She covered Irene's shattered knee with needles to kill the pain. We got her up onto her mule and she rode for two hours back to the rancho, followed by an hour's drive by van back to town, and then the overnight bus to Tijuana, and finally home. It was a forty-eight-hour journey with an injury that required surgery. She never cried out once thanks to the medical efforts of the lady from Israel and a few carefully chosen entreaties to the late friars of San Ignacio.

San Ignacio is home to no more than two thousand people spread out over a few miles. In the central square, you can purchase date pies from the lady who cut a hole in the wall of her living room to more easily sell them while she continues to watch her television. If you call for the only taxi in town, you will meet Gilbert, who looks like an accountant and often picks up his clients in coat and tie. Be forewarned that he will immediately bring out his puppets to make them dance, then segue into magic tricks like producing a coin from behind your ear.

It is not uncommon to see a vaquero ride his horse into town for supplies for his rancho in the suburbs, and on one very hot day, I watched one cowboy share an ice cream cone with his horse. Parking stalls around the town square are usually filled with camper vans carrying kayaks on their roofs. The local stores are packed with T-shirts that read, "I kissed a whale," and in the restaurants a delicious meal will set you back half of what it would in the States. Evenings, when the sun has finished baking, lights in the *cuadrado* (town square) will twinkle like fireflies and couples will appear, some promenading hand in hand. The hummingbirds, thick as fleas, will abandon the bougainvillea, and the cicadas will take over to sing through the night. The aroma of mole and tortillas fills the night air to accompany the beautiful melodies of a local guitar. If there is a band, as there often is, there will be dancing. San Ignacio is sophisticated beyond its appearance. The whales have made it an international hub where a dozen languages can be heard on any given day. They are the common link to everyone and the journey from town to the lagoon is sacred for many. Just as the padres of long ago traveled unused desert paths in their holy quest, so, too, do modern pilgrims travel the partially graded and pockmarked road that leads to the lagoon. It eats suspension systems, and I have passed numerous abandoned cars over the years. It took the visit of a Mexican president to pave at least a part of it. There is one final stop outside of town for a photo op. It's a large sign with a gray whale pointing the way that says, "60 km"—about thirty-five miles to go. After that, you enter God's land.

≈

As you leave San Ignacio to begin the thirty-five-mile (sixty-kilometer) drive to the lagoon, it becomes obvious why Spain never really conquered Baja. Their heavily armored soldiers on horseback baked in the desert sun and their horses succumbed to cactus wounds. They were no match for the guerrilla tactics of the indigenous people who could put an arrow in their back and instantly disappear into thin air in the vast expanse of desert. Instead of physically conquering the Cochimí with its military army, Spain sent a religious army that succeeded where soldiers could not. The mission padres forced the indigenous Cochimí to live on mission grounds where previously unknown diseases such as smallpox and measles from the old world wiped out much of the new. It was the same sad story of white contact told around the world, and by the late nineteenth and early twentieth centuries, these unfamiliar diseases condemned the Cochimí to extinction.

It was on one of my early visits to the lagoon that I met another of the town's more colorful denizens. My driver said his name was Taco. Thinking I had misunderstood him, I asked again, and he insisted that was what he wanted to be called. He was dressed in a tie-dyed T-shirt left over from the sixties and bright pink trousers under a wide-brimmed, straw cowboy hat, topped off with mirrored sunglasses right out of *Cool Hand Luke*. To make a point, nobody in town even gave him a second look. No more than a half mile west of town, we came upon what I considered to be a very large snake. It was probably four feet in length and was a dark orange mixed with brown spots. It was most likely a king or garter snake, but at the time I had never seen one like it. I asked Taco what it was, and he said he was not sure, so he casually pulled the van over to the snake and picked it up. It offered no resistance. Taco held it with two fingers, just behind the head as is the proper way to handle a poisonous snake. At this point the clients in the van were piling out with cameras and cell phones to record the moment. He brought it up to his face, nose to nose, as the snake's tongue flicked out to sense him. Taco then licked the top of the snake's head, declared it to be nonpoisonous with a big smile, and placed it back on the ground. I could only stare at him in amazement. I don't know if he meant it was not venomous, or was not poisonous to eat, but I did not have the heart to ask, and never did find out what kind of snake it was.

The next thirty-five miles is epic wilderness punctuated by tiny hidden ranchos. The land begins with rolling hills that open to vast dioramas of distant mountains and low, flat mesas. Much of the route is mineral-rich salt flats that, when filled with winter rain, turn deep shades of red, purple, and brown. From above they are a massive painter's palette. Several times along the way, you will spot water only to find it is a mirage, with the real thing still miles away. Ubiquitous turkey vultures soar overhead by the hundreds, always on patrol for the weak claimed by the remorseless desert.

The masters of this emptiness are the cardon cactus, cousins of the saguaros of Arizona that grow forty feet tall and live for centuries. When rain is scarce, the cardon shut down their growth process and begin to look like someone is tightening a belt around them, creating a waist. By counting the cactus's waists, you can tell which years it rained. Each cardon can weigh hundreds of pounds. Once expired, it leaves behind a strong, hardwood skeleton that is used for fencing. Cattle freely wander on this road, and it is not unheard of to run into a herd of sheep.

Most of the ranchos are tucked away from the road in arroyos and small canyons, their locations only betrayed by the smoke of wood fires. Any one of them would work nicely in a western movie. These are basic homes, built from whatever the desert has decided to give up, surrounded by fences of cardon skeletons with corrals filled with goats and sheep. If you stop, most of them are ready to offer fresh coffee and tortillas. No questions will be asked, and no criticisms offered. Money is usually refused because, to these people, it is simply the Christian thing to do. I used to have a favorite spot to stop with my clients where an ancient *abuela* would hand-knead the masa and cook tortillas in an outdoor adobe oven in the rear of the cabin. The last time I stopped there, the family was gone and all that remained was a small shrine with a photo of the *abuela*. Just as I mentioned along Highway 1, shrines and memorials abound in the desert. I have stopped at many over the years to wonder what story that person had to tell; how different their life must have been from my own to be buried by the side of a dirt road in a desert wilderness. The shrines are the physical manifestation of the Catholic faith that spreads throughout Baja, and tiny altars are a fixture in local houses as well; but the desert adds a touch of superstition. Usually, they consist of a small crucifix, some plastic flowers, and a sun-faded photo. If the deceased was a child, toys will be left behind, anything connected to the canceled life to show what they did while alive. If the family lives nearby, food is often added, and money is always left untouched by visitors.

Crossing a low, flat valley, the road ascends slowly, around blind curves dotted by occasional rockslides that often bury part of the way. There is one final, dramatic turn at the crest of a mesa that, when rounded, reveals a massive forest of cardon stretching for miles, like immense, silent sentinels. Their arms approximate human gestures that suggest a movie about the walking dead. Local lore says that at night, when the moon is full and the *brujos* fly, the souls of departed ancestors enter the cardon and walk for those who believe. More practical, the owls, kestrels, and crows pluck their thorns and hollow out a predator-safe residence.

By now your internal organs scream for relief and, thankfully, the road levels out to only partial washboard grade. The lagoon is near but first we pass hundreds of mounds of shells, each as tall as a human. At first sight they bring to mind burial mounds, and that is exactly what they are, but the deceased are all chocolate clams. The northern end of the lagoon is rich with these crustaceans, and for centuries they have been a staple part of the local diet. For many years, the fishing community dumped empty shells there out of convenience, and now, just as Northwest Coast peoples used crushed shells to make middens, so do the various whale camps collect and crush them to line the paths through their camps. They reflect moonlight well, and noisily let you know when someone is approaching. Coyotes hate walking on crushed shells.

At this point, navigation switches from landmarks to painted rocks or cactus with a tire around it. Homemade signs point down dirt paths declaring the various whale camps located there, and you wonder where that could possibly be. The land seems flat but there are tiny arroyos and washes to navigate through the spidery arms of thousands of blooming ocotillo. By now you are hungry, hot, sweaty, cramped, and tired, and you think the most terrible overnight journey you have ever taken cannot possibly continue, when the deepest blue appears as a thin strip on the horizon. It is the lagoon, but then again, it can't be. Is it another mirage? After endless hours of shades of brown, your eyes are suddenly assaulted with the most sublime, deep, rich shade of blue—a blue that God made on one of his better days.

FIVE

In the Lagoon

For most people, the first time they touch a whale is an intense, emotional experience. Many cry; some laugh hysterically. Irene and I both sobbed at our first contact. Interspecies contact with these giants brings out behaviors not usually expected from adults. Touching a whale makes us all children again and washes away all things bad. It is a moment that will never leave our memory and will become more sacred with each recalling. On one of my earliest trips to the lagoon, I was having a cup of morning coffee when a fisherman I did not yet know ran by, quickly followed by another. Without saying a word, I followed, not wishing to miss any action. I no longer recall their names, but they told me a baby whale had wandered too far from its mother and was trapped on a sandbar. We boarded a panga and found the baby quickly, cutting our engines to drift closer without further spooking the already panicked youngster. I judged it to be no more than a few weeks old, and a baby that young would not last long without its mother.

The baby was maybe twelve feet long and flopping around in water no more than three feet deep. The pangero told me to stay in the boat but I had to touch this unbelievable creature. Its skin was smooth and felt slick, as though it had a coating of silicone. It was unmarked by barnacles, and its pectoral flippers were flailing about in minor panic while it smacked its tail flukes over and over. The second pangero slipped into the water and covered the baby's head with a wet rag, then began talking to it, soothing it, caressing it like petting a dog, and I swear the whale calmed down immediately. They quickly ran a line around the calf's flukes that allowed us to tow it off the sandbar where not far away we found a distressed mother looking for her baby. The baby seemed to know it was being helped and did not put up any resistance. Back with its mother, we watched it attempt to breach, with great gusto but little success. The body language of both whales expressed their joy.

That was the second time I was witness to a fisherman talking to a whale, and both times they had spoken to them like a close friend. I had never known people like these who treated animals as equals, and its effect on me was profound. I thought of Padre de Porres and imagined him comforting the baby whale, and suddenly his story did not seem so implausible.

≈

Days on the water depend on the wind. If the tide is in, the pangeros will bring the boats right up to camp, and if it is out, we walk a quarter mile to a rocky point to board, past a nest built long ago on top of an abandoned *baño*, which brings the noisy wrath of a mother osprey who has newly hatched chicks beneath her. Near camp, the water is no more than three feet deep, and we can see the countless, tiny bat rays that resemble swimming tortillas with tails, scattering as our shadow passes overhead. While they may look harmless, they sting with a fiery pain, so no one swims because of them, and besides, the water is too frigid. The tidepools, captured by ancient basalt, are a universe unto themselves. At first look, they are merely colorful selections of inanimate sea life, but as your shadow moves over them, they come to life. Crabs no larger than a thumbnail scurry sideways, reacting to their natural fear of predators. Hermits, locked tight inside their shells, cover up and scurry along the bottom like prehistoric armored tanks. Purple and orange anemones open and close at our passing while our shuffling feet cause the sea grass to sway like a hula dancer. In pools no larger than a bathroom sink and no deeper than a water glass, there is endless life. Hungry gulls stand watch as we board the pangas. They will remain on watch until our return, never understanding why we are not bringing any fish with us. We watch our footing on the razor-sharp basalt, which can shred skin like broken glass, as we wade out to board. Dozens of tiny clams spit their displeasure at our passing. By panga, it is a ten-minute ride to the deep channels where we are authorized to watch the animals. Along the way, we pass low, flat, ancient mesas, built from centuries of storms depositing whatever detritus the sea has chosen to give up. The shore is thick with old whale bones that in time return to the sea as sun, wind, and water slowly break them down.

We float like a leaf on the current's mercy, the water's mirrored surface doubling the image of overhead clouds as great gray backs surround us in an ancient dance, the whoosh of each breath, a sigh of the Earth. Their raised tail flukes are a ballerina's jeté. A breach is the ultimate display of power and majesty, as for a brief moment, their forty-ton bulk approaches weightlessness. That feeling, in my opinion, is the main reason they breach. On mornings like that, I, too, feel weightless.

If the light is right when they blow, a rainbow prism, delicate as a dandelion, vanishes as quickly as it appears. For a calf in training, it is a game of hide-and-seek, trying to estimate where they will pop up and usually not where expected. They wear their scars like badges of honor, proof of their initiation into open ocean attacks.

Each season we also witness the tragedy of dead whales, called "floaters." They come into the lagoon on high tide and wash up on the beach, some with sides stove in from a ship strike, others simply dead from natural causes. Floaters set in motion an intricate chain of events in the pecking order of the food chain.

No sooner does a whale perish than Mother Nature sets to work parceling out its essence to those who need it. The gulls arrive first, covering the carcass in a squawking frenzy as they pick at the thick skin. They will have exclusive rights for a day or two and then the crabs will take over, stripping any leftover flesh with the dexterity of butchers. Gas quickly builds up inside the whale body, swelling it to twice its normal size, and the stench will reach as far as the wind carries it.

One day while investigating a dead whale, I had a client insist on having his photo taken with it. I warned him of the decomposition process and agreed to take the photo provided he did not actually touch the whale. Just as I raised the camera, he decided to lean against it and immediately found himself inside the world's largest pile of goo. He burned those clothes at camp, and everyone kept their distance for the duration of the trip because of his stench.

The position of lagoon warden rotates among the men of the fishermen's cooperative, giving that person full authority to issue fines for invading a whale's comfort zone. He waves at us from his beach chair at Rocky Point, the demarcation line that marks the official whale-watching channels, stationed to keep count of which boats are on the water, while counting and recording arrivals and departures of whales to and from the lagoon. A strict census of whales is maintained both coming and going that will be sent on to scientists to study and assure the continued migration.

We cross the lagoon to a beach littered with whale bones from past storms, but long before arriving, the aroma of thousands of cormorants and pelicans that occupy most of the beach becomes overwhelming. Two coyotes stand watch, hoping to find an infirm or elderly bird to make a meal but they will not attack a flock so large. We motor back to deeper water at low speed, passing Valentin in the rescue boat. He is an old friend that works for different camps, sort of the MacGyver of the lagoon. In passing he yells that one boat already has a friendly whale. We continue on until the pangero decides we are in a good place. There is much blowing around us so we cut our engines to let the breeze take us where it may. I can honestly say that during my years on the water I have never gone an entire day without making contact. I would also add that as much as I enjoy interacting with whales, I am just as satisfied to simply watch them. To me, they are moving works of art, mobiles, if you will, kinetic sculptures on a scale never imagined by Alexander Calder.

Even their barnacles can be attractive, like beauty marks on a woman. I look for clues to their lives, scars, prop cuts, the way they approach my boat. Each whale has its own personality, and in time I have learned to read their body language. Some whales always approach from the rear, probably attracted by the hum of the idling motor, while others insist on hanging out below the raised bow where it is all but impossible to see them. I have had mothers allow their calves to approach, but hung back themselves, and more than a few macho yearlings that like to wave their erect penises around in a game of whose is bigger. They do not always raise their flukes when sounding. If they are in shallow water, they will slide below the surface gradually, but if the water is deep, they will bring them up in a signature that says "look at me."

On the open ocean, the gray whale always leaves you wishing for more because all that most people will see is their blow and tail flukes. But in the lagoon, they display their soul to you, they say "come to me," they say "look at my baby," and they say "we forgive you for killing us." From Jonah to Moby Dick, whales have captured the human imagination, and while the secret lives of most aquatic species is hidden to us beneath the waves, the gray whale has chosen to share their secrets with humans.

All animals have a comfort zone, and we never invade it. They enjoy our presence and are there by choice, but they will only come in their own time. I will also add here that the lagoon is highly regulated as to how close boats may approach and how many may even be allowed on the water at any one time, so the animals are in very good hands. In January, the calves are tiny, still forming their limbs, and mothers are very skittish. Mothers will usually not approach a boat until early February, when the calves have filled out and are beginning to pursue their curiosity. Mother will give the panga a good look first, keeping herself between the boat and her charge. If she judges us safe, she will allow the calf to approach. This is the baby's reward for its morning training session. If it has done all that mother has asked of it, the calf will be allowed to come to us to play, but only for a short while. What we may view as entertainment is, in effect, a life and death training scenario that will determine if either mother or calf will return the following year.

Like a human infant, a baby gray whale has no depth perception, terrible balance, no common sense, and is nearsighted while out of the water, so they often crash directly and hard into our panga. They have the attention span of a gnat, surging from one side to the other, person to person, a pent-up ball of pure energy. They are desperate to play and are attracted to different noises and shiny objects. Music and singing have always brought whales to the boat. Newborns will circle us endlessly, bouncing around like a ball in a pachinko game, spitting water, splashing intentionally, and at times even lunging in an attempt to get in the boat. The calf has already forgotten its morning training session. It has done all that mother has asked of it, so the calf is allowed to come to us to play for a few precious minutes. The only proper description I can give to an encounter with a newborn is to imagine a rambunctious fifteen-hundred-pound puppy bouncing around from lap to lap. It is easy to forget they are wild animals and get caught up in the moment. I have had my hand crushed between a baby whale and the panga, and I have had clients who got smacked by a pectoral fin when they should not have tried to grab it. Being close to these frantically happy babies is a game of dodgeball, and we are usually the ball.

~

To me, the Pacific gray whale is a regal creature. It carries its own secret knowledge revealed only to those of us who take the time to learn about them. They do not kill their own, and they do not fight among themselves. They openly display affection and protect the weaker among them. They are as caring and nurturing as human mothers and as protective as human fathers. They communicate with each other and with us in their own way. Its current scientific handle (*Eschrichtius robustus*) was dispensed in 1864 in honor of a Danish zoologist named Daniel Eschricht. Its original scientific name was Latin, *Balaena gibbose*, which translates to "Humped whale." Since there was already a humpback whale species, this name for our gray whale is believed to refer to the knuckle-like humps, instead of a dorsal fin, that descend down its lower back.

The gray whale, after centuries of being hunted to the brink of extinction, has forgiven us and now comes in friendship. No other whale in its natural habitat approaches humans to be touched, and the same grays will not approach boats on the open ocean. The lagoon is their sanctuary, part of their DNA, and they know that humankind on the open ocean is a different species than in the lagoon. The lagoon carries a silent agreement between once ancient adversaries that says once inside its confines, we will not hurt one another. Such a covenant has made San Ignacio Lagoon a hallowed battleground and uniquely historic location of interspecies interaction.

No two gray whales look alike. Their appearance is as varied as that of all human cultures and their coloration differs just as much. Some have described gray whales as having been painted by Jackson Pollock with black, gray, and white paint. Some collect massive piles of barnacles, especially on their rostrums, while others may have very little. Their eyes face down and out, and their lids tend to droop. Combined with the downturn of their mouths just below each eye, it makes some people think they constantly look sad, and yet, while in a place like the lagoon, they are as happy as puppies. They are reminders that a frown is just a smile turned upside down.

Because they are the slowest swimming of all whales, averaging but three knots an hour, and they do not echolocate, they take refuge in the natural surroundings of the littoral, hiding in the kelp, holding their blow so as not to betray their location. They are adept at using rock formations to break up their underwater profile. I personally believe this shore-based proximity to people, in contrast to deepwater whale species, is part of why grays are so human friendly. I have seen them bodysurfing alongside people and, occasionally, curiosity will take them off course into a local marina.

While in estrus, females will mate with several males in a day, each time flushing out hundreds of gallons of sperm deposited by the previous suitor. The final doer of the deed becomes the father, who will exit their lives forever as soon as he is finished. Some of the calves will be born along the migratory route, while others wait for the luxury of entering the world in warm Mexican waters. Gestation is eleven months, and barring problems, a female will bear a calf every other year for up to four decades. At birth, the calves (pups are sea lions, calves are whales) are as bright pink as newborn orcas. They morph into their gray coating after only a few days and begin collecting barnacles soon after that. Mother will push the calf to the surface for its first breath because newborns come into the world with little instinct. At this point, they resemble crumpled beer cans with their fins and flippers wrinkled and twisted for a couple days, until swimming straightens them out. Whale aficionados call these little ones "pickles." Mother will only stop swimming long enough to give birth, then commence again, effectively dragging her newborn alongside in her slipstream since the calf is not yet ready to swim unaided. The new whale will enter the ocean at twelve to sixteen feet long (3.6–4.8 meters) and weigh close to one ton. It will reach sexual maturity at six to twelve years, and as an adult, its heart will weigh 250 pounds and be the size of a wooden barrel. If it evades predators and harpoons, it can expect to live fifty to seventy years, while one female was documented as living almost ninety years.

Gray whales must be taught everything from breathing to eating. Once in the lagoon, mother will have six to eight weeks to prepare her offspring for the marathon swim back north, educating in a real-time classroom. In their multi-mile marathon swim, the grays must evade their natural enemies, orcas and sharks, every day. Because of their slow pace, many bear the scars of combat. They are favored by small sharks who take tiny bites of flesh the whale never feels through its coat of six to eight inches of blubber. The calf will nurse for about seven months, daily consuming three hundred gallons of mother's milk, which is 50 percent fat, while gaining as much as a hundred pounds each day. (Human milk weighs in at 3.5 percent fat, so, fortunately, we humans cannot gain a hundred pounds a day.) The calf does not attach to a teat, but instead, mother squirts her milk into the water from her genital slit for the calf to lick up like a thick gelatin. Mother whale will not eat while she is nursing and will lose up to a third of her total body weight. If the mother dies during that time, the calf will most likely die because gray whales are not known to adopt orphans as other whale species sometimes do. Once the calf is weaned, mother and child will usually part ways for the rest of their lives, but on some occasions I have seen yearlings still with their mother. Grays rarely travel in pods, and when they do it is usually a small one. Having said that, I once witnessed a pod of fifteen grays off Dana Point, California, on their sway south. Normally, they are solitary animals, but the natural world is always in flux.

The majority of whales in the lagoon are females with their young. Males tend to hang outside the mouth of the lagoon or in the southernmost area. Why they do this is uncertain, but I believe it is because they do not want to interfere with the infant training, or just can't be bothered by it, like some fathers. Males are not known to show affection to newborns. Occasionally, gang-like pods of immature males will bully their way around, waving their erect penises in the air in a macho display of whose is larger. The gray whale has even endured charges of homosexuality because, for reasons unknown, males will sometimes huddle with their enormous penises entwined, but there is no actual evidence to support this as a theory of sexuality. It should be noted here that the gray whale penis can be up to six feet long, varies in color from soft pink to bright red, and is prehensile. During the throes of passion, it sometimes waves about like a flag in search of a female's genital slit. More than once, my panga has been whacked in the amorous and sometimes violent act of mating. The gray whale penis is an unintentional weapon to watch out for.

≈

Approaching boats is a learned behavior. If the mother whale approaches boats, it means her mother and grandmother did also, and she will teach her offspring to do the same. But if she does not approach boats, the baby never will. The baby learns by imitating its mother, and here is a tender example of that. A mother whale with calf approached us one morning. Mother came in first, slowly, making sure we were no threat. Her tentative approach told me perhaps she was a first-time mother and so, rightfully, cautious. She mouthed the gunnels of our panga, running her mouth most of its length while we on board stroked her flanks. She then backed off, and eyeing us from a short distance, waited a few seconds for her calf to approach. When it did not, she gave it a gentle nudge and, using her rostrum, pushed it alongside our panga just as she had done. The baby grew very excited at this new sensation, and I felt extremely gifted to have witnessed a learning moment as the newborn whale was introduced to both boats and people.

≈

Out of about two hundred to four hundred whales that enter San Ignacio Lagoon annually, perhaps only twenty-five will be "friendly" whales. When you consider the worldwide reputation of the lagoon, it is fascinating to note that it is based on a minute fraction of all the whales that occupy the entire rim of fire. It is that very small minority that makes the lagoon unique. In January, mothers are overprotective of their charges. By February, the youngsters have grown enough that mother will gradually let down her guard and allow them to come to us. March is my favorite time, because the calves are now in full training mode, curious and energetic. Besides, mom now needs some relaxation after her grueling training sessions. That is when both mothers and calves begin to come to us in numbers. By April, many have already left the lagoon to head north. When mid-April arrives, only the friendliest whales remain. Some may hang around as long as May, reluctant to leave their newfound human friends.

Since the purpose of the lagoon is to act as a nursery, there is a daily routine for them to follow. Youngsters must prepare for the return swim north, so their mothers make them swim back and forth endlessly through long channels in the center of the lagoon each day to build stamina. Like all newborns, baby whales tire quickly and are distracted by anything that moves. Even a breaking wave will distract them. Just as human children in school need a recess to let off steam and take a break from learning, so do baby whales. That is when mom will bring them to our boat to be petted as a reward. Sometimes she will hold the baby on her back or roll upside down with the calf on her stomach, and sometimes a mother will hold her charge aloft with a pectoral fin. I am convinced that this is not only a reward for the youngster's good training period, but a chance for the mothers to proudly show off their handiwork: "Hey, look what I made!" Nothing is funnier than watching a newborn whale trying to climb on mother's back, sliding down one side and then the other, and splashing with rambunctious energy. Baby whales are curious, playful, lack depth perception, and can on occasion be very obnoxious, such as when they willingly slam their flukes on the water to create a wave next to the panga, or deliberately spit water at us. They will ram head-on into our boats, and even lunge, trying to get in with us. They can act just as silly as any small human child. One youngster came shooting straight up into Irene's outstretched arms for the briefest of moments in what would have been a perfect hug had she been ready. That was a missed photo she and I will always regret. We decided later it was attracted by the shiny silver bracelet she had on her wrist. They watch us as much as we watch them. For the whales, we are mere bathtub toys, only there for their enjoyment, but that is a role I do not mind playing.

During all this, mother is close by, usually logging on the surface only a few yards away. For a mother to place her calf between humans and herself is the ultimate gesture of trust, but mothers need love also. When she decides that baby has had enough, she will push it aside and come in for her own one-on-one time. On occasion, both mother and calf will push and shove each other to rule our attention. On more than one occasion, we have had as many as five adults around our boat, all pushing and shoving, having a great time at our expense while their calves watched, hoping for their turns at the boat. I have had my arm in a whale's mouth numerous times, often up to my shoulder, scratching a bright pink tongue while mother holds perfectly still, enjoying the sensation. They also like to have their baleen stroked. For creatures lacking hands and feet, touch is a most precious commodity.

Americans more than any cultural group seem to possess an inbred need to touch things they should not, and while we are on the water to do exactly that, there are some rules. I tell the clients to never touch the eye, never grab a flipper or tail, and most important, never put one's finger into the blowhole. That may sound obvious, but people try to do it all the time, so I have to inform them of the powerful valve inside the blowhole that closes when the whale dives. The valve is so strong that if a human finger were in it, it would sever the finger or drag the person into the water. To date, I have not lost a finger or a client to a blowhole.

Next, there is kissing. Most whale watchers in the lagoon are not content to simply touch a whale; many want to kiss one, and the whales often cooperate. I am amazed how many times over the years a whale will hang motionless next to the panga while people take selfies and try to kiss them. That is when I tell the client about the three distinct species of lice the grays carry inside their barnacles. The barnacles themselves are white, and the yellow color comes from the lice that have taken up residence inside them. For hygienic purposes, I do not recommend kissing a whale, but confess to having done so many times myself. After all, it is a rather unique club. People do not hesitate to make a fool of themselves when it comes to kissing a whale for a selfie. For those who doubt their friendliness, try such a feat with a lion in the wild.

This is a good time to talk about barnacles, because they go together with gray whales like chocolate and peanut butter. A barnacle is a tiny crustacean that is related to crabs and lobsters. There are almost fifteen hundred different types, and they float freely in the ocean attaching themselves to stationery and slow-moving objects such as sailboats and gray whales. They attach by glands that secrete a quick-drying cement that scientists have studied in hopes of artificially producing it for commercial use. They secrete four to eight plates of calcite to form a protective shell that is razor sharp and resembles tiny volcanic cones. They are filter feeders, taking in minute plankton as they pass through it while riding on their host. On slow-moving animals like the gray whale, they often form giant clumps on their rostrums, sometimes as large as a basketball. When the whale rubs a barnacle off, it takes skin pigment with it and that is why older whales have white faces much like elderly dogs and white-haired people. Many believe whales breach to rid themselves of barnacles, but barnacles can withstand hundreds of pounds of pressure and are difficult to dislodge. They are found the world over and are a fact of life for many sea creatures.

Barnacles on gray whales carry three species of lice unique to themselves, and one of these is appropriately named *scammoni* for the whaler who once hunted their hosts. In a side note, the US Navy has speculated that the buildup of barnacles can slow the speed of a military ship by up to 60 percent!

Gray whales have been blessed with a unique sleeping habit they share with a few other remote creatures. Mothers will only let their pups sleep as long as they estimate it might take a predator to reach them from their horizon. How do they make the calculation? Well, they eyeball it, but it's tricky because their eyes are on the sides of their head and angled down and out. While underwater, gray whales have excellent vision, but above surface, they are nearsighted. So, mother will spy hop; if it is shallow, she will stand on her tail to facilitate this. She can then see what is immediately around her while also estimating the time it would take a predator to reach her—and that is how long she will allow her calf to sleep. Seals and walrus that haul out on the arctic ice employ this same sleeping method using their nearly infallible inner clock.

During the migration both south and north, grays swim nonstop, employing the same unihemispheric slow-wave sleep technique as orcas by shutting down half of their brain while the other half remains alert, but much of the time, it is predator estimation that dictates sleep. Since they must breathe air to survive, a part of them must always remain conscious to prevent drowning.

Grays also have excellent hearing. Like true seals, they have no exterior ear, but only a tiny hole behind each eye. They vocalize in registers too low for most people to hear, but according to Dr. Aaron Thode of Scripps Institute, who began an acoustic project with the whales of San Ignacio in 2005, sound travels five times farther underwater than in the air, allowing whales to communicate with one another over vast distances. Discovering this information took me back to the words of Owen Chase and the harpooned sperm whale that Chase said he could hear. He was in a dory at the time, some distance away from the mother ship, and the sound could have skimmed over the water to him, but it is more likely that the death throes of a dying whale are much louder than everyday vocalizations. Hydrophones in the water can often pick up the sound of whales speaking with one another. I should add here that the largest of all whales, the blue whale, has such a low-frequency voice that it can speak to one of its own kind in an ocean on the other side of the Earth.

In the 1980s, cetacean researcher Marilyn Dahlheim, a biologist at National Marine Fisheries Service, began recording the vocalizations of gray whales and believed that they make six major sounds, including moans, belches, clicks, knocks, and an unusual metallic noise they send out as a single pulse. Most humans have an eight-octave hearing range, but it is believed that gray whales have a twelve-octave range. I can say without reservation that gray whales react to human vocal encouragement. When we yell and cheer, they become visually agitated in the absolute best ways. They understand they are putting on a show, and being natural hams, they love it. They also have a sense of humor that manifests itself most often with water being spit directly at our cameras.

≈

A mother whale has four to six weeks to build baby's stamina for the swim to the Bering Sea, so most of the day is spent endlessly swimming laps, but what inner voice tells her when the calf is ready? Only a mother's instinct. As the calves mature, mothers will move a bit closer each day to the *boca*, the mouth of the lagoon, getting the baby used to the heavier surf and strong tidal exchanges before venturing out into the open ocean. The tide flows quickly in the lagoon, and it is known for creating rogue waves. It would not take much near the *boca* for a baby to flounder. If the surf is mild, mother may take her calf outside the mouth into the open ocean for brief periods of exploration before returning to the lagoon.

During the training session, I just sit back and watch. Mother might begin with a lesson in lunging, a technique whales employ when running into a bait ball on the surface. Mother will make a dramatic lunge, pushing walls of water ahead of her with mouth wide open and lower jaw fully extended, then baby will make an anemic copy that is more of a flop than a lunge. For the first several tries, they usually make it halfway out of the water before falling over to one side like a toppled bowling pin. All regular gray whale behaviors must be practiced endlessly before the baby is seaworthy. Spy hopping is a bit easier and the one most calves master first. I have been fortunate to get one photo after all these years of a mother and baby spy hopping simultaneously. Once a baby learns to properly spy hop, it often adds a spin to it, turning around and around like an aquatic top. Breaching is the tough one, reserved for calves a bit older. Mother will circle underwater, building speed to propel her forty tons up and out of the water. They can sometimes get as much as three-quarters of their body out of the water, but when you weigh so much, that is a hard order to fill. I have only seen one whale breach completely free of the water and that was a humpback with the largest pectoral fins of all whales. A gray whale has tiny pectoral fins and must generate speed with its tail alone. A baby's first few breaches are nothing more than lunges because it takes time for their pectoral flippers to become strong enough to aid their tails, but once they catch on, it is not unusual for them to breach dozens of times. I have observed a personal record of thirty-seven breaches by a youngster. For reasons beyond my knowledge, for most that I have seen, adults breach sideways while youngsters tend to turn upside down while in the air. I remember one young whale that breached directly next to our panga. I had enough sense to yell for everyone to cover and I stuffed my camera under my jacket just as a massive wall of water showered us. Everyone loved it until the whale breached a second time with the same results. Anticipating a third breach, everyone had their cameras ready when the whale came up on the opposite side of the boat, thoroughly drenching all of us and our cameras.

~

Gray whales have baleen instead of teeth. Baleen is keratin, similar to human fingernails. A gray whale has up to 160 baleen "filter plates" that hang from the top of its mouth. There are none on its lower jaw, and these plates are how it filters food from the water. The baleen near the front of the mouth is short, and it lengthens as it recedes back inside the mouth. While the whale is young, the baleen is soft and as pliable as cotton, but as it ages, the baleen yellows and turns brittle just like human fingernails. Grays are bottom feeders, and they are right- or left-sided just as humans are right- or left-handed. They will cruise along the ocean floor, scooping up tons of silt and sand along with the minute crustaceans that live in it. They will push the sand and excess water out with their tongue and lick the remaining crustaceans from the back of their baleen to eat. We know from necropsies that whales prefer one side over the other as the baleen on one side of the mouth is usually worn down much closer to the jaw, and often the eye on the preferred side is opaque and blind from years of rubbing on the sand.

Its tongue will weigh as much as three thousand pounds (1,300 kilograms), and a gray can maneuver it to dislodge food stuck in its baleen.

One day we were in shallow water watching a mother whale teach her calf how to feed off the ocean floor. The water was clear at first but quickly became pure muck as the mother cruised along the bottom sucking up tons of silt and water. Most grays I have observed lunge after bottom feeding to push the excess water out with their tongue in a massive purge of mud and silt. You do not want to be near them when they do. On occasion, they will make a sound while expelling the water that approximates a large animal evacuating its bowels. That day, we could not see three feet through the murk when the baby exploded out of the water right next to us in a perfect imitation of mother, and vomited gallons of ocean bottom muck that coated us from head to toe.

Because they are warm-blooded, grays require an immense daily intake of calories to maintain their core temperature of 99 degrees F (36.5 degrees C), taking in 4 percent of their body weight, or about 2,400 pounds of krill and amphipods per day. That intake amount is becoming increasingly more difficult to maintain, a problem we'll address later in the book.

On occasion they are opportunistic feeders, especially if they run into a bait ball near the surface. While they are not considered to be rorqual whales like the humpback (those with expandable lower jaws due to multiple throat grooves), the gray has three to five grooves on its lower jaw that allow it to expand it in much the same way as a pelican does. Babies will often test this expandable lower jaw by filling it with water and go about spitting. While submerged, babies also like to blow bubbles that explode at the surface. The pangeros jokingly refer to this as a whale fart.

There was one youngster I will call "Bubbles," because it circled our boat several times before surfacing, blowing bubbles as it went. In a few minutes, our panga was surrounded by bubbles popping on the surface. When the baby finally showed itself, it raced around the panga trying to swallow the bubbles. Then, like a fountain, it leaned backward and sprayed the boat with a lengthy stream in a very deliberate act.

≈

We have shared this planet together for eons, but many people have no idea just how connected we are to these animals. Like humans, whales are born with a soft spot on top of the skull known as a fontanelle, to allow for expansion of the brain. We are both warm-blooded, air-breathing mammals that give live birth, nurse our young, and have hair. In the case of whales, they are singular hairs on their rostrums that act as sensory devices, picking up movement in the water around them. Whales have a vestigial pelvic bone from when they walked the Earth some six million years ago. In the womb, a human baby has flippers that morph into individual digits before birth, while a whale fetus has individual digits inside the womb that morph into flippers before birth. A whale has every single bone from its shoulder to the tip of its flipper that you and I have. The first biblical reference to whales appears in Genesis, "and God created the great whales." Thirty-seven additional whale references follow in the Bible. The Icelandic sagas tell of giant sea monsters that breathed fire from the tops of their heads. That was the first literary reference to a whale blowing. Our connection to these animals is long and deep. Some whales, like the humpback, have proven their species to be altruistic many times over, often coming to the aid of divers by placing themselves between a shark and a diver. Several videos exist of humpbacks physically returning distressed divers to their boats. Unfortunately, grays do not seem to have this quality, but they make up for it in pure friendliness. Even though we have hunted whales for centuries, they continue to return to us in friendship.

And what about grays' relationships with other whales? Well, there must always be a negative yin to a positive yang, and for the gray whale it is the orca. Baby gray whale is the preferred food of the orca. Some biologists have speculated that mother whales can hold off giving birth until they are in the safe confines of the lagoon, but when coming and going, baby grays are in constant peril from the planet's mightiest predator. The orcas will attack from the rear, often in a pack in a verbally coordinated hunt. They will try to sever mother's spinal cord near her flukes to take away her only real weapon. Should she manage to strike an orca with her tail she can deal it a good blow, but in the end, she will lose the fight. Mother will always impose herself between her calf and its attacker. She will try to get the baby up onto her back or roll over to place it on her stomach, hoping to sacrifice herself instead. I have seen footage of a mother gray whale in her death throes, smacking an orca with her tail flukes hard enough that it ceased its attack, but by then, it was too late. She will not stop fighting until her final breath in the purest expression of a mother's love.

As an apex predator, the orca prefers to drown the baby by climbing on top of it and holding it under. The tongue will be eaten first, a killer whale delicacy, then the rest of the carcass depending on how hungry it is. If the orca is well fed, it will discard the rest of the carcass. It is nature at its bloodiest and most violent. What is left after an attack like this will be quickly consumed by sharks and lesser predators. The skeleton of the whale will sink to the bottom, called a "whalefall." Within days, it will become home to organic growths and small critters will feed on it, so even in death, the whale plays a role in the perpetuation of the ocean.

Six

Maldo Fischer, Johnny Friday, and Frank Fischer

If Pachico Mayoral is the godfather of the lagoon, then the *alcalde* is Romualdo Liera Fischer, known to one and all as Maldo. The Fischer name is everywhere throughout San Ignacio, and Maldo is aware of the position he holds. He is a "man of honor" and beloved by everyone. His colorful background conveys a brief history of San Ignacio, and it goes like this: In 1910, Maldo's grandfather Frank Fischer was the German fourth mate on a merchant marine vessel docked at the French colonial mining town of Santa Rosalia on the Baja coast of the Sea of Cortez, several hours' drive southeast of the lagoon in the early cars of that time. His ship had put ashore to pick up copper ore. The twenty-five-year-old Frank got into a ruckus with the vessel's second mate and thought the looming punishment serious enough to jump ship. The story of what the fight was about has several versions, so the reader can use imagination here. Now a fugitive, Frank headed overland and hid out in the mountains until things cooled down. Eventually he made his way to the village of San Ignacio where, being handy with tools, he quickly became the town's only blacksmith and eventually married a local girl. He was good at what he did, and no one was interested in his past. Before long, he was considered the premier mechanic on the central Baja coast with a reputation for keeping a car running, no matter its condition. This was at a time when there were no roads, and the ones that existed barely qualified as such. Because of his international maritime experiences, he spoke several languages, which came in handy, and he made most of his own tools. Mechanics were in short supply at the time, and Frank quickly became a local legend. He was mentioned in a 1958 article in a small circulation publication titled *Solo Below, a Guide to Lower California*. In 1971 he made it into the much larger circulation *Sunset* magazine, and a 1974 guidebook titled *Guide to Baja* also included a reference to him. Frank was the undisputed car doctor of central Baja and resident celebrity of San Ignacio, and he was lucky no one from the merchant marines read *Sunset* magazine.

Most important, Frank unwittingly played a role in the preservation of some of Baja's most precious artifacts. While on a hunting trip, he discovered the first of what would prove to be approximately four hundred painted caves and rock shelters in the area. That cave had an enormous snake painted on the wall, and today it is known as "Serpent Cave." In the mid-eighteenth century, there were a few references to cave paintings in the journals of Jesuit missionaries, most notably by Padre Joseph Maxiáno Rotheax, who apparently visited some of them. In 1789, Javier Clavigero published an account of some of the caves in his *Historia de la Antigua ó Baja California*. His was the first assumption that the pigment used to paint the caves came from the Three Virgins volcanoes. He also wrote that some of the enormous paintings, many of which were more than twenty feet tall, proved they were created by a race of giants, fueling yet one more legend to attract adventurers and explorers.

In the mid-nineteenth century, an amateur anthropologist and naturalist named Leon Diguet was working in the French colonial town of Santa Rosalia mining copper when he got wind of painted caves in the area. It was the same Santa Rosalia from which Frank Fischer would flee years later. Diguet led a mostly French expedition into the mountains in 1893–1894, and a year later he published the first known scientific report of painted caves in the area. Whether Frank read that article and already knew of the caves, no one knows, but it was Frank who got credit for discovering the Serpent Cave. In April 1962, the writer Erle Stanley Gardner, of Perry Mason fame, fancied himself an amateur Baja explorer. When he heard of this Serpent Cave, he arrived by helicopter and hired Frank for a tour. Realizing there were more cave sites to be explored, Gardner returned many times, bringing a photographer from *Life* magazine and an archaeologist from UCLA named Clement Meighan, who after visiting several different sites published the first scientific article about them in a journal. Eventually, Serpent Cave made its way into Gardner's book *The Hidden Heart of Baja*, and an additional article he wrote for *Life* magazine. It was not until the mid-1970s that a Baja explorer named Harry Crosby, assisted by a local guide named Tacho Arce, went searching for the original route of the Camino Real and found and documented two hundred separate painted caves. It did not take long for the paintings to be considered as significant as the paleolithic paintings of Europe. Among those two hundred caves were the images of whales left to us by the Cochimí. Thanks to Crosby's discovery and documentation of these paintings, in 1992, then Mexican president Carlos Salinas de Gortari inaugurated the largest archaeological project in state history to study these early people and their art. In 1997, Harry Crosby published *Cave Paintings of Baja California*. Crosby remains one of Baja's best-known explorers.

Today, while the Fischer name is known throughout central Baja, few people remember or give credit to Frank for being a pioneer in the preservation of some of the peninsula's finest works of art, but they do know his descendants. Maldo, one of Frank's grandsons, was born in the local mountains, most likely under the care of a midwife, as that was the norm at the time. His family still keeps a ranch south of the lagoon called Rancho Baleena. He made his living as a fisherman, as did most of the men of San Ignacio at the time. He married a local lady named Catalina Murillo, a San Ignacio native who bore him two sons. Maldo had a good life between fishing and taking the occasional whale watcher out on the water.

Fast-forward to 1993, when a thirty-four-year-old surfer dude from Southern California by way of Texas named Johnny Friday was camping near Maldo's home and going out to see the whales with him. At that time, there were no organized whale-watching camps, just individual fishermen taking the occasional curious client out on the water. Most of the fishermen were rather poor and used gill nets to maximize their catch, which often included young whales that got snared. Johnny and Maldo pooled their money to buy two nylon tents from Walmart and started a humble camp that gained immediate popularity. There was no actual road from town to the lagoon at the time. Directions were "turn right at the big cactus and left at the painted rock, and don't get stuck on the salt flats or you will die."

In those days, reaching the lagoon was like traveling to Timbuktu. Johnny and Maldo used the income from their first clients to buy a used army surplus tent to cook in, and that was the beginning of what was then called the Baja Adventure Company. It was 1998 when we all met on my first trip to the lagoon. During an evening in which I consumed too much brandy, I painted a small sign with a whale on it welcoming people to Campo Cortez, and for years, that was the official sign for the camp. The camp name was chosen to honor the early Spanish explorer and carried a certain panache as it rolled off the tongue. Today, Campo Cortez is the queen of several whale-watching camps, and it has won numerous environmental awards from the Mexican government.

My first trip to the lagoon began after finding the company during an internet search. Having encountered an orca in British Columbia the previous year, I wanted to learn about gray whales next. Irene and I had already made a disappointing journey to Magdalena Bay in 1997 where we saw lots of whales, but none approached us. It was overcrowded with boats, and they were harassing the whales, invading their comfort zone. I was put off by the entire experience. I found out later that Mag Bay was not under the umbrella of the biosphere and so was poorly regulated. I did my homework and decided to visit San Ignacio Lagoon the following year. Since then, I have wondered if that choice was a conscious decision or fate taking my hand.

Irene and I flew into Loreto, Baja, where Johnny Friday had agreed to meet us at the airport and drive us to the lagoon. I was impressed that the owner of the company himself was going to pick us up, unaware that he and Maldo, whom I had not yet met, were the *entire* company. When we stepped off the plane to an empty terminal, we took a taxi into town to the information center where no one had ever heard of Johnny Friday. I was furious! Irene and I had paid our money, flown down to Mexico, and were stranded in an unfamiliar place. By that time, I had formed a foul image of Mister Johnny Friday; I envisioned this sleazy conman with slicked-back hair and shirt open to the waist with gold chains around his neck, gleefully counting ill-earned money. I had been suckered and was ready to search all of Baja to find this feral fellow and retrieve my cash.

Just as I was relating to myself all the horrible things I would do when I found him, a blond, blue-eyed surfer dude, looking like he'd just left his board outside, rushed into the tourist office out of breath, asking if we were his clients. After profuse apologies and fragile explanations (he had asked a friend to pick us up while he gathered our supplies, but the friend forgot about us), we piled into a funky old travel van and shared the back seat with an enormous propane bottle for the six-hour ride to the lagoon.

That was how I met Johnny Friday twenty-five years ago, and I spent the next twenty seasons working as guide/naturalist for what began as the Baja Adventure Company and evolved into Baja Ecotours. Today, almost three decades later, Baja Ecotours receives the most clients of any camp in the lagoon and continues to win environmental awards. Maldo has taken over whale-watching operations while Johnny spends most of his time in La Paz where he has become a major underwater cinematographer with his own film company, Baja Productions.

≈

Here's a story about five fools on a panga. Irene and I plus two friends on their first visit to the lagoon; Maldo was driving our boat. We noticed a lot of splashing and slowly motored over to the churning water, spotting a pectoral fin and half of a tail fluke protruding from the water, indicating a female whale on its side. This is a position they assume to secrete their milk into the water while the calf closes in and licks it up with its tongue. That is what we thought was happening. As we drew slightly closer, Maldo uttered a loud, "Oh, no!" and threw the panga into a hard turn.

Nursing often resembles mating, both of which may take place on the surface.

We had inadvertently approached a mating *trio* of one dominant male doing the act, while a second, beta male, hung below supporting the female. This is a common mating practice that can become quite violent. Sometimes rival suitors will line up and begin pushing and shoving for their turn, and without warning, the sea can become a churning quagmire. At this point, the dominant male's Pink Floyd, as it is commonly referred to, was standing as erect as a flagpole. When a male gray whale is sexually excited, his penis becomes a lethal weapon and he flings it about with little regard to what is nearby. Maldo backed us off several yards, and we sat silently, waiting to see what came next. After several minutes of total silence, our female boatmate peered over the side of the panga to see the large male, directly below us, on his side, looking right at her! She inched back into the panga, silently gesturing that the whale was directly below us, just hanging there. A couple more minutes passed, and she cautiously looked over the side again. At that moment, the sky turned black as his immense tail flukes came up hard, whacking the side of the panga and deluging us with a waterfall. It missed our friend's nose by inches as we rocked back and forth. It could easily have sunk us but chose to give us a little tap as a warning. We were soaked but still had a good laugh at our own expense. I made a note to learn more about the nuances of both mating and nursing, and since then I have wondered if any whale I have met was the product of that interrupted mating. I have also thought of the great dinner conversation that could be had about how a whale broke a friend's nose.

SEVEN

Campo Cortez

There are few places on this Earth where reality transcends expectations and the environs can honestly be called mystical. The lagoon is more than just a place; it is a feeling, a state of mind. A friend once defined the feeling as "duende," a word that has no literal translation. To me it means soul, essence, and a lust for life all wrapped into one word. Duende touches one's soul. San Ignacio and Campo Cortez are the world before graffiti and pollution, before mechanical noise and skies filled with contrails. There is no tapping on a keyboard or FaceTiming on an iPhone. If you stop moving to listen, you will hear only the rhythm of nature in this place where time exists only when necessary. One does not enter a place like this without purpose. For some, it is an escape, while for others it is a quest, and there are always those who go out of simple curiosity. It is not a place for a vacation because that would cheapen it. It is a place of learning and connection where we go to feed our souls. I have never defined what calls me to such places, but I have always been grateful for the call.

Campo Cortez, my part-time home for two decades, is named for the already mentioned Spanish explorer Hernán Cortés, who brought most of what was then pre-Mexico under Spanish rule in the late sixteenth century. It is a tidy, self-contained village of twenty wooden cabins, all solar powered with comfortable beds, hot indoor showers, flush marine toilets, and a communal dining *palapa* (an open-sided structure typically with a thatched roof) that drips ambiance along with delicious meals. It sits at the edge of the Earth on layers of igneous rock embedded with countless shells of creatures who lived under the sea eons ago. The land feels old, as if nothing would be different if we stepped back centuries in time. The family that runs the camp lives in private homes about five miles away, on a tiny bluff overlooking the northern part of the lagoon.

Long before the camp was truly organized, we stayed in heavy canvas tents that the nightly wind battered loudly. It was like sleeping inside paper bags. The beds were surplus army cots, and we wore layers of fleece to keep from freezing. There was one large surplus tent in which the entire social life of camp took place—cooking, eating, and gathering at night to discuss the day's events. For me, I had wandered into a way of life I had never considered and found myself liking it. Meals were concocted on a single propane stove, and water was stored in large industrial containers. Showers then consisted of plastic solar bags with drain nozzles left to heat in the sun, then hung from a clothesline. Bees loved those shower bags and covered them like a natural hive.

In time, camp evolved with the addition of two-person wooden cabins, solar powered by car batteries. Today each cabin has two permanent beds with firm mattresses and bedding, sliding windows, and storage space, plus a battery-charging station. The local ladies adorned each cabin with whale-print curtains. Pathways of crushed clam shells have been laid to preserve the local fauna, and there are tiny solar-powered lights throughout camp. There are two enclosed showers with a flash heater. Because fresh water must be trucked in weekly, we ask guests to shower every other day, which is not a problem when you and everyone around you smells like a whale. The only real inconvenience is the location of the *baños*. For sanitary reasons, they are about fifty yards from the closest cabin, which at night in a frigid wind can be a brutal walk. The resident coyotes often take up station a few yards away to watch the nightly exodus to the *baño* as their evening entertainment. That led to us filling a shelf with hospital bedpans and plastic buckets to use inside the cabins at night, to be emptied in the flush toilets the following morning. In a thick morning fog, the line of people bearing buckets appears like a scene from *The Walking Dead* TV series.

The current camp is entirely the work of Maldo, the renaissance man. He and his crew built and wired the cabins and dining *palapa*, all solar- and wind-powered. As far as wilderness camping goes, it is quite luxurious, and the entire camp is nomadic, making its own annual round-trip migration. Whale-watching season begins in early January and terminates by the end of April. After that, camp is broken down, piece by piece like an immense jigsaw puzzle, and every item regardless of size is shipped south to Johnny Friday's compound in La Paz. Everything from the windows to electrical wiring, the bathrooms, and showers, leaving only the gutted cabins and communal *palapa*. The reason for breaking down and storing the camp is that it is several miles from where the Campo crew live yearlong, and they will be out fishing after whale season, unable to keep an eye on things until the following year. In early December, everything is trucked back, and the entire camp is reassembled in time to greet the first whales.

The Campo is a family operation. Everyone has a job from the smallest toddler to Maldo in command. For the children, it is not work because each day combines play with learning. As soon as they can walk, they begin to interact with people from all over the world and take pride in being part of the larger organization. Because they are so remote, the children attended school by video long before Zoom became part of our technological landscape, but their real classroom was all around them. The smallest child knows every critter that lives above or below ground and most of the birds passing overhead. They can discern poisonous animals and plants, and can tell an animal from its track in the sand. From an early age, they are in tune with their surroundings, reverential toward all living creatures. They learn the animals' stories from their fathers and absorb empathy from their mothers. From the time they are swaddled infants, they are taken on the water to meet the whales, because they have been born into a family whose role it is to care for them. Because of the international clientele that visits the camp, most of them are bilingual if not multilingual. What they may lack in material things is more than compensated by the wonders of the natural world. Any child should be counted fortunate to grow up in such an environment.

At Campo Cortez, the day begins at first light. The crunch of the pangeros' gum boots on the crushed shell paths awakens everyone in their cabins. Maldo's sons, Roberto and Paco, will already be in their daily uniform of orange slicker and buff, as will their uncle, Cuco, all of whom will take us out onto the water. They are on their way to wipe down the boats and fuel them for the morning's whale watch, but first they will shoo away the dozen or so pelicans that have roosted on board for the night where coyotes cannot get to them. As nature awakens, Catalina and her sister Elsa will be on their way to the cooking *palapa* where they will don their aprons, which feature a smiling whale in a chef's hat, to brew strong, fresh coffee and begin making tortillas for breakfast by hand.

Let's visit one particular cabin, a nondescript little building that most people assume is a tool shed. We perpetuate that tool shed myth. The interior is no more than six feet by six feet. On the side that faces away from camp toward the lagoon there is a picture window. On the roof there is a satellite dish, and inside there is a bed, easy chair, and flat-screen television. This is Maldo's refuge. When he needs time from the everyday running of a large operation, he retires here to relax with no interruptions. Once inside we do not bother him except for the most dire emergencies. No client has ever asked me much about that tool shed, and I have never offered to tell what it really holds. It is an inner sanctum not to be violated.

When the sun first breaks the horizon outside my cabin, the water turns to liquid gold, and I watch expanding circles on the surface, ripples created by jumping fish, interlock as they spread before me. With the camp now alert, the pelicans will begin their daily hunt, making kamikaze, headfirst dives to take fish. To keep their necks from breaking and their trachea from tearing, they turn their heads slightly to the left at the final moment before impact. It is a move so subtle that you would have to study them for days to notice it. Caspian terns are also common with their pointed wings slicing the air like knives. Like their pelican cousins, they are lunge feeders and can spot a fish below the surface from a half-mile away. Once they go into their power dive, the fish has no chance. Their counterparts are the goofy-acting red herons who stick their entire head into the sand to probe with their long, curved beaks for the fast-burrowing sand crabs. Willets and whimbrels, balanced on impossibly thin legs, peck at minute aquatic creatures too small for the human eye, using their curved beaks to ferret them out of the mucky sand. The numerous species of sandpipers all have different length beaks thanks to the forethought of Mother Nature and evolution. All pipers probe for tiny creatures in the sand, and the different lengths of their beaks assures that they will all eat from a different sand layer. Therefore, they never compete with one another for food at one particular depth. Plovers run back and forth, just ahead of the tiny waves generated by the morning winds, pecking like jackhammers in the split second before the next wave slides in. Since they nest directly on the sand, we are always mindful of the fragility of their tiny homes and even more delicate eggs. Behind the dining *palapa,* there is a small sandy beach that occasionally fills with tens of thousands of pipers, and when they decide to take wing en masse, they nearly blot out the sun, twisting and turning in an aerial ballet that would mystify a jet fighter pilot. All of these birds that frequent the littoral are a study in perpetual motion and possess blazing high metabolisms, so most of their day is spent foraging.

Of the approximate three hundred bird species that inhabit the lagoon, my heart goes to the osprey. *Pandion haliaetus* is a large sea hawk or eagle, and just as the whales rule the water, the osprey rules the sky. They are a single species found worldwide near bodies of water with a sufficient food supply and are as common as sparrows in the lagoon. They are diurnal raptors whose diet is almost entirely fish, and they are as skilled at catching them as their human counterparts. They will soar at a couple hundred feet, spotting a fish more than a mile away. They swoop in with a last-second flair of their wings, dancing on the sky, extending their talons and scooping up prey in a single choreographed moment. Ospreys occupy a single nest their entire life unless it is destroyed. It can weigh several hundred pounds and be large enough for an adult to lie down in it. Over the years, I have spent days watching the males bring tree branches to the females who painstakingly weave them together with the skill of an engineer. At one time or another, every cabin in camp had an osprey nest on top, so several nesting poles were erected around the site. They are all in use, but it did not stop ospreys from building on top of the cabins. Oddly enough, there are no trees within several miles of our camp and yet the male ospreys are always bringing in large branches, so they must travel long distances to find appropriate limbs for their homes. After my morning osprey fix, I look around for fresh coyote tracks, because the trickster makes a nightly patrol for anything left outside. He is not fussy, as it is the act of taking he craves. I cannot count how many shoes I have found on my walks through the ice plant (*Delosperma*), a spongy plant that grows like weeds and surrounds our camp.

Post-storm mornings are the time for beachcombing. Neptune deposits both trash and treasures along the shore in the aftermath of his wrath. I follow the sandy path north through ice plant groundcover, past the *baño* in the guest camping area, with the roof caved in from the weight of an osprey nest. I follow the low ridge of ancient basalt embedded with millions of shells, once home to living creatures for millennia before humans appeared on the planet. There is no step without a shell underfoot, and my sandals crunch those not yet petrified. I am walking over a massive crustacean graveyard and know the land has changed little since these creatures lived. A few yards out, my feet sink to the ankles in the soft sand of low tide, and the cold water makes me feel alive. I imagine what sort of dinosaur stood where I do now and try to imagine the land eons ago.

The mornings when the fog lays thick are my favorite. The haze dissolves the present, and its cottony void gives me a feeling of weightlessness, as though I can take wing like a bird. The silence lets my mind wander to the heroic days when mother whales fought to the death to defend their babies. I can stand at water's edge and stare into the void without regard for time, connected to the place and its history. All that has passed since then has conspired to bring me to the very same shore. Who knows how many whales died in this aquatic battleground that is now a sanctuary? If I listen, I can hear the cry of a harpooned whale and the curses of the whalers chasing it. I watch the water turn crimson until I realize I am staring at a patch of kelp.

A half-mile from camp is our "boneyard," a massive array of cetacean bones that, now residing on a beach off by itself, once sat in the center of camp as a teaching aid for visitors. There are whale ribs longer than a human, compression discs, vertebrae, scapula, and even a fully articulated dolphin skull. It was a fine miniature natural history museum of value, but that all ended one day when a gentleman in a suit appeared. When Maldo told me this story, he added that he always had his guard up when anyone arrived in a suit. The government official looked over the collection of bones and declared that Maldo was operating a museum without a permit. Maldo tried not to laugh, thinking at first that the official was joking. He was not. He gave Maldo a timeframe to apply for a permit or remove the bones. That was the man's way of asking for money and that is also why they now sit on the beach a half-mile from camp, clearly visible, but out of the government's jurisdiction. Now they have become an outdoor classroom and are popular with the camp guests.

The most fragile inhabitants of the lagoon often wash ashore after a storm. Sea horses leave behind an exoskeleton of bony plates, and oftentimes their eyes solidify and remain intact. I think them beautiful in both life and death and have collected dozens over the years to show the children in camp. I have found everything from glass fishing floats to plastic buoys, nature-sculpted sea glass, and even a fairly intact pair of Oakley sunglasses along the shore. The sunbaked bodies of puffer fish and skates are a common sight, both toasted to a mummy-like consistency by Mother Sol. On these walks I usually find coyote beds from the previous evening; the sea grass will be tamped down or the ice plant crushed. The coyote will circle and nest just like a dog, until the grass, or in most cases, ice plant, is just the right consistency. They rarely stay in the same place two nights in a row, and I can tell what their dinner was from the bone pile they inevitably leave. Where I live most of the year in a city, coyotes are considered to be a problem, but in the lagoon, they are an intricate piece of the ecological puzzle that keeps the food chain intact. Being scavengers, they eat just about anything from rabbits to berries, fruits, and vegetables, even rattlesnakes if they threaten their den. During a hunt they can reach 43 mph (68 km). Most important for Campo Cortez, they eat rodents that would overrun the camp if unchecked. They are solitary hunters but will gather in a pack, called a "band," to take down large game. Their hearing and sense of smell are excellent. They can hear a twig snap from a half-mile away and have superior night vision.

There are nineteen subspecies of coyote. They can mate with wolves and dogs, and they are monogamous. If a mate is trapped or injured, the other will not desert it, even at the cost of its own life, and if trapped with no help, they will gnaw off a limb to escape. In indigenous lore, there are countless stories about the coyote, more so than any other animal, and they are often referred to as the "trickster" because of their wily intelligence.

There is no definitive evidence as to why coyotes "howl at the moon." They have at least eleven means of communication, and howling is a way to tell others of their location. I like to think that just as whales breach for fun and because they can, so do coyotes howl at the moon. In camp they are so used to people that I am constantly telling clients not to touch them.

≈

Everyone, guests and crew, will gather for breakfast. There are always fresh, hot tortillas, rice and beans, but for the more health conscious, there is yogurt, cereal, and fresh fruit. We check the weather report and tidal chart and decide if the boats can go out or if Mother Nature will confine us to camp for the day. If it is too windy, we may meander through the mangrove channels to watch birds. From the window, I look west to barely make out the morning blows of several whales in the channel that will take us about fifteen minutes to reach by panga. By 9:00 a.m. we walk to the point, a basalt finger that sticks into the lagoon with its own layer of ancient shells—our launch point during low tide. The resident oyster catchers will walk away at our approach, looking back at us all the while, too dignified to fly, and mother osprey sitting on her nest above the *baño* will loudly announce that we are too close to her domain.

As we put into the water, I watch dull brown rays, no bigger than tortillas, slide along the bottom, running from our shadow, and I count the hermit crabs migrating for the same reason. A tiny puffer fish cruises with just the tip of its dorsal out of the water. The overhead cloud canopy will assure a mild morning. Paco will lead the way out, because Paco only knows one way to drive, pedal to the metal in a bone-jarring ride to the whale channels. A chop on the water only means that Paco must go faster to fly higher over it. The rest of us just smile and roll our eyes while we hang on. Roberto and Cuco will pilot additional boats, and if we are lucky, Maldo will also drive. He does not go out as often as in past years because camp has grown and requires his almost constant attention. If he goes, I will be in his boat, hoping that his whale magic will bring them to us.

A full breach lifts forty tons halfway out of the water

A spy hopper standing on its tail to have a look around

A baby whale circles the boat and spits water at us

Coming to the boat to have its tongue scratched

Inverted gray whale uses its tail flukes to catch the wind and travel

(ABOVE) Two large females wanting to be petted

IGHT) Getting a nose scratch and having its baleen stroked

Kissing a whale—
not sanitary, but a popular event

Author petting a baby gray

Baby whale twisting upside down during a full breach

Two females aid a newborn onto its mother's back

A mother nuzzles her newborn calf

Baby asleep on mother's back

Pachico Mayoral, the godfather of the lagoon and the first known person to pet the "devilfish" (*Getty Images*)

Paco radioing on the water

(ABOVE) Valentin, the lagoon's jack-of-all-trades, and the author

(LEFT) Author poses with Johnny Friday and Maldo Fischer, the founders of Campo Cortez

Maldo and Catalina, the behind-the-scenes "queen" of the lagoon, with the author welcoming visitors to camp

(ABOVE) Campo Cortez in the afternoon

(RIGHT) Jim and Irene at the lagoon

People find their release from daily life in many ways. For me, freedom is an open boat on the water. That moment we push off is like the spreading of wings. To cruise at speed in a panga is to fly above the Earth. There are no problems on the water, no wars, no fires, no shootings—there is only the moment and the whales. I become ancient and insignificant, and yet I am important. I have a sacred trust to educate those on the water with me, and in that moment, I realize I have the best job in the world.

As we pick up speed, dozens of cormorants floating on the water begin their crazy dance routine of taking off. They flap their short wings for lift and literally "run" over the water, their tiny, webbed feet flopping like clown shoes. They cannot simply take off as, say, an osprey can but must get up to speed by running over the surface, wings flapping just enough to seem as though they are sliding before they lift off into the air. As we pass them, they spread their wings wide to wave at us. One of the most ancient and efficient fishing birds, cormorants generate great speed underwater and can maneuver quickly enough to chase a fish. Underwater they can dive to 150 feet (approximately 45 meters). For that reason, fishermen have employed them for centuries, especially in Asia where their owner places a rubber ring around their throat that prevents the bird from swallowing the fish it catches. The bird returns to its master on a tether, and once it regurgitates its catch, it will receive a reward of a fish tidbit.

When we return after a whale watch, I like to find silence at the end of the tiny peninsula that holds Campo Cortez. It is basalt surrounded by ice plant and is subjected to fierce winds coming off the open ocean. If I am still, within minutes I will be joined by pelicans and cormorants, all of us looking out to sea to ask our various private wishes. It is my peaceful time at the end of a busy day.

Dinners are always fascinating and not only because of the food, but also because of the wet and smelly people we have spent the day with. On any given night there may be multiple languages in play around the table, and as the cerveza loosens tongues, the stories begin to flow. I have had wonderful conversations with scientists, poets, artists, philosophers, filmmakers, and even a politician or two. The camp has even hosted two Mexican presidents, but I was not present for either of them. Many of Mexico's most prominent literati have visited camp at one time or another, and it is they who have consistently given public support to keeping the lagoon safe and pristine. Almost every author who has been a visitor has left a signed copy of a book for the library, which boasts close to a hundred volumes. Two of my own books reside there.

On my most recent visit, I met Cheryl and Nick Dean, independent filmmakers whose latest offering, titled *The Witness Is a Whale*, won the 2021 International Outdoor Film Festival Grand Prize. There at Campo Cortez, we had a private screening on our flat screen in the dining *palapa* for twelve people, the first time the film had been shown outside of a film festival and prior to its official premiere at the Explorers Club in New York City. The film is a scathing exposé of secret and illegal Russian whaling operations.

As Mother Sol gently sinks below the horizon, we gather lawn chairs and face west to watch the sky become a kaleidoscope—yellow turns to orange, then red, and finally to a soft magenta. All the great artists in the world could not mix a palette equal to that which coats our camp each evening. The sun sets directly over a perch of an osprey family, turning them into a perfect silhouette as the day's final gentle hues fade into darkness. When there are no clouds, every star in the heavens shows itself. Under Campo Cortez skies, I learned most of the visible constellations, hidden by the lights of the city where I live the rest of the year. During whale season, the Big Dipper stands directly over the dining *palapa*, pointing my way home with the North Star. The Little Dipper is never far away, and Orion's Belt is a regular sight. Moonless nights are the most dramatic as the Milky Way looks as though a giant hand has cast tons of glitter across the universe, and if you watch straight up and do not focus, you can clearly see the International Space Station as it sails through the cosmos.

It is then, during high tide, that I lay on my bunk and listen to the splash of jumping fish outside my cabin, and if the wind is right, I can barely pick up the blow of a whale in the far channel. Under a full moon, I can hear mournful coyote arias and listen for the crunch of shells as they send their first night scouts into camp to forage. The night sky is ruled by the great horned owl whose baffled feathers allow it to attack silently from behind. I sleep deeply in the lagoon, untroubled by the worries of my personal world back home, happy to be free of television news and the internet, as I fall asleep to the tune of nature.

Come morning, I check my water shoes for scorpions and the front porch for rattlesnakes. In the shallows, a red egret is poking its head into the sand in search of sand worms. Behind the dining *palapa*, tens of thousands of willets fill the shore, the same color as the sand, blending in and seemingly causing the beach to dance with movement. I head to the *palapa* for a cup of Elsa's coffee. It is time to begin another day on the water.

EIGHT

The Campo Cortez Pangeros

I have known Paco and Roberto since they were little boys, their uncle Cuco keeping them in line. Both have grown into men of respect with their own families now, assuring that Campo Cortez will continue to be run by the Fischer family, who cannot imagine doing anything else. Like many who share his given name, Francisco Javier Fischer Murillo is called Paco by everyone. He was born in Bahia Asuncion in lower Baja. His brother, Roberto, is Jose Roberto Fischer Murillo, and was born in the lagoon at home where Maldo and Catalina still live in an area known as Campo la Base, which only has a few homes. In Mexico, males often take the name of their mother to follow that of their father; thus the Fischer Murillo names.

Paco and Roberto are opposites; Paco is short and stocky. He has a profound sense of humor but is often shy. Roberto is tall, lean, and stoic in manner, speaking only when he has something significant to say. Both share an encyclopedic knowledge of sea life and are as at home in a panga as on land. Paco drives with a lead foot, while Roberto is more laid back and patient. Each day on the water with either of them is a learning experience as they pass along nuanced information as only those who share their lives with the whales can. They have taught me how to read an animal's body language, to anticipate what they might do next, and how to position the panga so it does not intimidate the animal. They taught me to scan the water without focusing to catch the flash of white and light gray just before a whale surfaces. They are always conscious of the importance of photos and strive to keep the sun on the animal rather than presenting it as a silhouette. With only endless water for reference, they always know exactly where they are in the lagoon, even out of sight of land. They know where it is deep, where it is shallow, and the animal behaviors they expect to see in each place. They know by the feel of the wind on their skin and the taste of the spray in their mouths. They know how the weather will change from the pattern the wind makes on the water's surface, and they can taste rain days before it arrives. Like nomads negotiating a desert of sand, the pangeros navigate the immensity of the lagoon as their forebears taught them. In the vastness of the water, where shore can be only a thin line on the horizon, they are at home. Maldo once told me long ago that he may not always know exactly where he is, but he is never lost.

Since they are fishermen most of the year, the pangeros taught me many of the habits of the other denizens of the lagoon, like bottlenose dolphin, puffer fish, multiple shark species, sea lions, and once, a very young *mola mola*, more commonly known as a sunfish, the largest of all boned fishes that can reach 5,000 pounds (2,268 kilograms). It was the pangeros who taught me about the octopus, among the smartest creatures on Earth, whose brain runs all the way through their tentacles, allowing independent movement of each suction disk. They can unscrew jar tops, open door latches, and slip through an opening no thicker than a piece of paper, and I found that if you pick them up when they do not want to be handled, you get squirted with some serious purple ink. The pangeros taught me to find octopus homes by looking for the tiny wall of rocks and shells they build around the den opening to deter predators. If you stick a finger inside, they are so curious they will grab it and you can pull them out for a mutual look at each other. Paco used to drop concrete wall blocks a few yards out from camp that quickly become octopus condominiums. Paco and Roberto also taught me about the great sea turtles, especially the giant leatherback who can weigh anywhere between 500 and 2,200 pounds (249 to 998 kilograms), and whose armor shell can reach seven feet from tip to tail. Of the seven known species of sea turtle, you can find six of them on the coast of Baja. In the heat of summer, green turtles will come ashore to lay up to two hundred eggs at a time, always at the exact spot where they themselves came into the world, and if only one makes it to adulthood, it is a major success story. Sea turtles are plentiful on the Baja coast. In sail-and-cloth whaling days it was customary for whalers to pick them up because they offered no resistance other than flapping flippers as they continued to attempt the motion of swimming when lifted from the water. They were allowed free run of the deck as sailors' pets, but more important, because they live more than a year without food or water, they were always an available food source during a difficult journey. Today, San Ignacio Lagoon is a principal feeding area for the giant leatherback, hawksbills, green, and olive ridley turtles. Many enter the northern areas of the lagoon to lay their eggs when the whales leave.

One of my favorite stories regarding Paco happened on a beautiful day when we were near the *boca*, or mouth, of the lagoon where I thought the surf to be treacherous. To Paco it was just another day at the office. We usually only venture that far when there are no whales anyplace else. On that day I was the naturalist for a boat full of people I did not know. We spotted a mother and calf and headed in their direction. For reasons I cannot explain, I looked back at Paco on the tiller just as a massive wave was approaching from our stern. We call them rogues and they are occasionally generated by wind and currents converging in a specific spot.

This one was huge and coming fast.

Not wishing to frighten the clients, I poked Paco and motioned for him to look back. When he did, his eyes grew as large as grapefruits. He had lost track of how close we were to the *boca* and hit the gas hard. The panga lurched with a neck-snapping jolt, and we took off with the people raising their hands as though riding a roller coaster, not realizing we were about to be swamped by a rogue wave. Paco gave it all he had, and when we were out of danger, the clients applauded, thinking he had deliberately given them a little thrill. He and I remain the only ones who knew how close we had been to an unwanted bath.

And so it has been, all these years, in our communication with the whales. The pangeros and I have come to know individual whales personally. On a day when Paco was driving, we spotted a large female with a white birthmark on her right flank, the size of a pizza, which allowed us to track her. I called her "Patch" because of her birthmark, but she stays in my mind for a different reason. She liked to come alongside us and just stick the top eight or ten inches of her rostrum out of the water for me to scratch her chin. She was surprisingly free of barnacles, but had an open sore on her rostrum that looked like she had spent a lot of time rubbing it on boat keels. She would press against my hand, and she did this each time I saw her over three seasons. She was so docile, so aware of our fragility, she gave off an aura of tranquility. I believe she recognized me and returned on purpose when she saw me. Her snow-white rostrum told me she was quite old at the time, so I consider myself blessed to have known her when I did. I have interacted with so many whales over the years, one might think it impossible to remember individuals, but I believe there is some sort of chemical reaction between interspecies friends, just as there is among people. There are some whales you just take to immediately, for no particular reason, and others that fade from memory.

A few years back, Maldo kept a pet Chihuahua named Skipper who had the run of the camp and liked to bark at and chase coyotes who usually just ignored him. As the coyotes grew more accustomed to humans, they became more aggressive, and one night we heard a mournful scream that chilled our bones. We rushed out of our cabins to see a coyote trotting right through camp with Skipper hanging from its mouth by the scruff of his neck. Like a shot, Paco took off in pursuit at a speed I did not think him capable of, and with a diving lunge that would have been a great football tackle, he grabbed Skipper from the thoroughly shocked coyote who kept on going. Unfortunately, a year later, Skipper chased a coyote into the ice plant and was never seen again.

When Paco is not driving a panga, chasing down raiding coyotes, or dodging rogue waves, he etches glass. His artistic heart creates delicate underwater scenes populated by the creatures of the lagoon. It is an art form that requires the touch of a surgeon and patience of Job, and I have several of his pieces in my home.

My favorite Roberto story took place on another beautifully clear but windy day. We were dead in the water watching a whale sailing with its tail flukes in the air. This is not common, but occasionally a whale will do this, and move about by wind power. I had four clients on board, none of whom had even been on the ocean before, let alone seen a whale. Suddenly a shark about six feet long popped up and began to circle the boat, occasionally raising its head to stare at us with its cold, black eye. Neither Roberto nor I gave it much thought as sharks this size are common in the lagoon and pose no threat to the whales other than an occasional cookie-cutter bite that they do not feel through their eight-inch blubber. Sharks are also not in the habit of attacking boats three times their size, but I could tell the clients were a bit nervous.

Now, all pangas carry a bailing bucket, just in case. I pointed at the shark and gave Roberto a look. With an economy of movement, he grabbed the bucket and bonked the shark on its snout, its most sensitive part, and the shark took off as if shot from a gun. Roberto replaced the bucket and resumed his usual statuesque stance at the tiller. The clients, who had their backs to us, never knew what happened. I just told them that Roberto had made the shark go away, and they may have believed me. I have had clients so naïve of the ocean, they believed anything I told them, so have always been careful with my words, but in this case, he really did make it go away. Roberto's face turned upward with a satisfied grin.

In his spare time, Roberto carves wood, but that is an understatement. When you create an object with pleasure and ease, it is an art form. One afternoon while we sat casually talking in the dining *palapa*, he picked up a small block of wood and began whittling with a pocketknife. I paid little attention to what he was doing, being immersed in our conversation. Within minutes he handed me a beautifully carved gray whale, with detail I thought not possible from such a basic tool. Roberto has the soul of an artist, and wood conforms to his talent as easily as most of us breathe. Today, that whale, bearing Roberto's signature, resides on my desk, and it is a prized possession.

More recently, Roberto has revealed his philosophical side with his postings on social media about family and friends. He is one of those people who goes quietly through life, drawing little attention to himself, but accomplishing important things. Both he and his brother are worthy successors to their father.

On my best day out with Cuco, he introduced me to a whale he referred to as his personal friend—a large female who approached us from the stern. She came directly up to the rear, where Cuco was at the tiller, and spy hopped him. He spoke to her in a soft voice, while stroking her rostrum, as one might address a child. He laughingly referred to the whale as his "girlfriend," because he said every time he saw her, she would come to him, and I gave him a hard time about having "whale magnetism." The whale stayed several minutes, and another boat was waiting to come in and make contact, so we backed off to another spot. Within minutes, the "girlfriend" was at our stern once again. No matter how many other boats were around us, that whale followed Cuco like a lost puppy. When it was our time to return to camp, she followed us for miles. The next morning, the same whale was the first one to approach us. While many of the pangeros know individual whales from so much contact, that was the first time I was convinced that this whale really was a "friend," as much as a forty-ton wild animal could be.

When he is not driving a panga or the "ride to live bus," Cuco plays drums in a Tejano band and often entertains at the camps. He has a very laconic manner and often speaks under his breath. When few others can hear him, that is when I listen most intensely. He comes across as quite world weary, which may seem strange for one who lives in such an isolated place. But the people of the lagoon are very international. Most people travel to experience and learn, and while the good folks of Campo Cortez do not travel much internationally, the lagoon is such a mystical place that it brings the world to them. It is travel in reverse, and the pangeros are as knowledgeable of the outside world as any who travel through it.

Cuco probably knows more about whales than anyone except Maldo. He taught me that when the tide goes out, the whales spy hop with increased frequency, most likely to make sure they will not be stranded in water too shallow. He also thinks they like the feeling of scratching their tale flukes on the bottom. It was Cuco who taught me about "crowd behavior"—when a group of whales, not necessarily in a single pod, all exhibit the same behavior simultaneously. Some days, dozens of whales will spy hop, popping up and down like jack-in-the-boxes, while on another day they may breach en masse. I recall one day when a half dozen different whales came to our panga with open mouths. Of all the behaviors I have witnessed, opening their mouths is probably the least practiced, and yet there are days when every whale in sight wants its tongue scratched, or its baleen rubbed. It is as though they all get together and decide on a specific behavior to practice. Cuco thinks they may associate a particular behavior with a particular panga or even an individual pangero. If that is the case, I would take that as proof of their cognitive abilities and well-evolved memory. I wish each of the pangeros would author a book, because collectively, they could give the world more information about whales than any classroom-educated biologist.

Valentin holds a special place in my heart. He was the second person Irene and I met at Campo Cortez after Maldo. We had just arrived, hot, dirty, and tired, and collapsed on our cots when a sprite of a young girl unzipped the tent fly and scooted underneath one of our cots, taking both of us by surprise. When I looked below to see what she was doing, she held a finger to her mouth, telling me to be silent. A second later an even smaller tot let himself in and, upon seeing what I then took to be his sister hiding under Irene's cot, he ran out of the tent. A minute later, with the diminutive fugitive still hiding, the mustachioed visage of Valentin poked his head in, introduced himself in English, and—I remember this moment as if it were yesterday, because Valentin has one of the world's all-time greatest smiles—he asked if we had seen his daughter, and we both played dumb. At that moment, the tiny girl began to giggle, and the jig was up. Jennifer was Valentin's daughter, and the boy who made a quick cameo was his son, Gadriel, and together they were playing hide-and-seek. By the end of that day, Gadriel had made an elaborate drawing of a gray whale that he gave to me. There was no shyness in these tykes. Clients are constantly coming and going at camp, and I thought what a beautiful life they have, with no inhibitions regarding people they do not know.

For the next twenty years, regardless of his age, Gadriel presented me with one of his whale drawings every trip, and I have kept every one. Now that I am no longer working there, I plan to return all of them to him as a precious memory. Today, Jennifer is married with her own daughter and is embarrassed when I tell her she once hid under our bunks.

Valentin is a dear friend and a man of many talents in the lagoon. Like a pinch hitter coming off the bench, he works at various camps doing whatever job needs to be done at the moment, driving pangas or sometimes the search-and-rescue boat. He is also an accomplished hookah diver, a trade that fascinates me. A hookah rig is simply a full-face mask with a long breathing hose attached. The hose runs up to a compressor on a panga on the surface that must be manned by a second diver to assure a continued air flow to the man below. The depth is limited to the length of the hose. The diver walks along the bottom with weighted shoes, digging up scallops and placing them in a net basket to be hauled on board. On the murky lagoon bottom, with water constantly stirred by passing whales, visibility is limited at best. I asked Valentin once if he had ever run into a whale while down there and he laughed, saying, "No, only a couple of sharks." I have no idea if he was joking or not.

Years ago, Valentin's wife, CeCe, was our camp cook and I got to see their children often. Gadriel had an insatiable curiosity for all living things in the lagoon, and I remember him bringing tiny critters to show me—one time it was a horned toad and another time he caught a tiny snake. He would make up stories about these creatures, sometimes assigning them voices, and together we would go on adventures with his critters. His best move was to come into my tent one day and take off his hat to set free a large frog that hopped onto my cot. Another favorite Gadriel story took place when he could not have been more than six or seven. I walked into the dining *palapa* to find him sitting at a chess board playing with a friend. I knew how to play chess and was impressed that they had set up the pieces correctly. His friend asked me to play, and I asked Gadriel if he minded. He shook his head and said with a smile, "Watch out for him." I took my seat and his friend asked if I was ready. When I said yes, he picked up his king and knocked my king over, declaring, "Checkmate," and with that he walked outside.

I remember a day in Valentin's panga years ago when a mother whale approached with her calf. At first, it was a typical encounter of petting and stroking, but when a second panga appeared, the youngster grew agitated and began racing back and forth between our two boats. Each panga kept its station, not attempting to get any closer to the whales, and it was the baby that repeatedly went from boat to boat. It would slam headfirst into the side of a panga, then turn to do the same to the opposite boat. We were not sure if the baby was playing, but it was using such force that we feared it might injure itself and we broke off contact. That baby followed us for more than a mile, chased by its mother, whom we were sure was trying to rein it in but without success. That was the only time I have seen such troubling behavior and it made me realize that perhaps whales can have mental distress just like humans.

The ladies, who mostly stay behind the scenes, are an integral and equal part of the camp. They operate the place like an efficient resort. Catalina and Elsa run a kitchen worthy of a Michelin star, and to do so at the edge of the wilderness makes it all the more remarkable. Meals are varied and delicious, with lots of fresh salads, whose ingredients are trucked in weekly from town. The seafood is to die for. Desserts are baked in a propane oven. Almost every group of clients I have had in the lagoon have asked Catalina to show them how to make tortillas, and she always obliges. Besides the scallops and clams we dig ourselves, we frequently have lobster, sea bass, grouper, and an occasional halibut. Catalina's chicken mole is the best I have ever tasted, and both ladies make wonderful seafood casseroles that always bring calls for seconds. On top of that, they do all the laundry as clients constantly come and go, all the while making sure each cabin has clean sheets, pillowcases, and ironed curtains. While she never acts like it, Catalina is the queen of the lagoon to Maldo's king and is loved and respected equally.

In her spare time, Catalina sews wonderful little fabric whales, and many of the women in the lagoon create arts and crafts from whatever local detritus there is. They make carved statues from tree limbs, they paint bean pods, string seashell necklaces, and make whale tail earrings. Some of them operate a small store at one of the camps that also sells T-shirts, coffee mugs, and more. Even in the middle of nowhere, people want souvenirs of their journey, and all the local works being organic, with the money from the sales staying within the lagoon community, they are popular with visitors. On weekends some of the women shuttle their wares from camp to camp, and I confess that my favorite piece has become a breaching whale napkin holder that has a place of honor in my kitchen.

The pangeros brought this book to life. They are modern caballeros, horsemen, and epic figures as much as the mounted vaqueros. They are the connectors between us land-dwelling creatures and our finned, watery friends. While I learned my cetacean facts and figures in a classroom at an aquarium, it was the pangeros who made it all real for me. I have been on the water on the calmest days and during a tempest sea, but never in fear when a pangero was at the helm. Watching them at the tiller is a study in concentration, head constantly swiveling, watching all directions at once for the approach of a whale, feet spread wide for balance and a smile of satisfaction on their lips. Their work is their life, and they are as much a part of the lagoon as the whales. On days when the whales do not come, I cherish our conversations with only the wind and water listening. Their English is far better than my Spanish, but neither is necessary to communicate. We use hand signals and nuances learned from countless hours in a panga together, our gestures having become a personal shorthand. But for all of us, a silent day on the water is a magnificent story.

I confess that it is impossible to encapsulate the lives of such people in a few short stories. Words could never pay proper tribute to even one of their normal days. The crew of Campo Cortez are at once ancient warriors and modern pioneers. They have foresworn much of the conveniences of the modern world for a simple but epic life devoted to protecting nature and its children.

≈

It was one of those perfect mornings when I met a whale with a crooked mouth. Either she had a broken jaw that had healed improperly, or she was born that way, but regardless, her lower jaw did not line up with the top of her mouth, so she always looked like she was sneering, and her right-side baleen was always visible. She came right up to the boat, to me in particular. Sometimes a whale will pick a person and stick with that one human all day regardless of who is in the panga. I named this whale Slackjaw. For three days, as soon as we arrived in the whale channels, Slackjaw would be at our side, nuzzling like an old friend, and I had to wonder if she was just a very friendly whale or if she recognized me from a previous year. She might have known me before her injury. Though damaged in appearance, she was one of the sweetest creatures I had met on the water, nuzzling as close as possible, staying right with me when we moved the boat. She was like an oversized puppy that did not want to leave its master. Because of her deformity I did not feel right about stroking her baleen, so I just ran my hand along her jawline, and she would always open her mouth wider as I did so, convincing me that it was a smile. She seemed to almost sigh as I did this. On a good day she allowed me to scratch her tongue, and she seemed to like me rubbing the tip of her rostrum. She found our panga three days in a row and spent most of each morning with us, only leaving when our time on the water was over, and even then, she would follow us back to camp until it was too shallow for her to continue. Other than her mouth, she appeared to be in good health, but I never saw her again after those three days.

It was Slackjaw who made me think about the possibility of whales adopting people with a look, just as dogs do. I am not alone in the belief that dogs hug people with their eyes. I know my own dog does this. That is why when I am with a whale on the water, I attempt to look directly in one eye, and most of the time, they return the stare. Simple as it may sound, I always hope that one day a whale will surface and speak to me in a language I can understand, but until then, it is enough to look into their eyes and know inside what matters.

On one of my favorite days, I was in Cuco's panga when a large female approached. She logged directly next to us, rolled to one side, and opened her mouth slightly. It was just enough for me to run my fingers along her jawline and stroke her baleen. That was the first time I had touched baleen in a living whale, and it felt like the softest cotton. I never gave a thought to her closing her mouth and trapping my fingers. She was totally docile and obviously enjoying the sensation. She hung there, unmoving, as I casually stroked her over and over. After a minute, Cuco called my attention to two additional whales that were approaching. As they neared the panga, they, too, opened their mouths, and both Cuco and I were terribly busy for several minutes, stroking their baleen. I looked up for just a moment to see three additional whales heading our way, all of them with mouths already open. They were like a line of puppies following a trail of biscuits, all six whales surrounding our panga demanding to have their baleen stroked. It was no coincidence, and proof of the crowd behavior Cuco had told me about. The first whale must have told the others that she had found a boat where the people would stroke their baleen, and word spread quickly. All six whales hung in one spot while we raced from one whale to the next, laughing like schoolboys and having the greatest time. It was one of those moments when you wished everyone you ever knew could see what you were doing at that very second. We lost track of time and were only brought back to the present when the radio crackled, reminding us that our time on the water was up. I am sure most people do not believe me when I say I have spent hours with my arm inside the mouth of a whale, but I have the photos to prove it.

On a typical morning on the water, we followed in Paco's wake, bouncing like Ping-Pong balls in a washing machine and yelling with joy. At the rocky point we cut our engines and waved at the warden seated in his lawn chair on the bluff, counting the boats, allowing only sixteen at a time in the whale channels, and making sure no one was entering their comfort zones. We were allotted two hours at a time on the water, and when one boat came in, another was allowed to go out, but everyone knew that if we had made good contact, we would stay out longer. It is at this point that I often see whales spy hopping by standing on their tails in the shallows so they can people-watch those who are watching them.

I've always preferred the morning, when the wind still sleeps and the surface is flat. Blows spouted like countless lawn sprinklers all around us so we cut our engines and drifted, knowing it would not be long until one of our gigantic friends approached. I looked back at Maldo, who was smiling and waiting to work his magic.

I have always considered Maldo to be a mentor, even though I have quite a few years on him. He has saltwater in his veins and the world's biggest heart, which he opens whenever a whale approaches him. He calls petting a whale "talking with fingers," and he does such a perfect imitation of a sea lion that the pinnipeds (seals, sea lions, and walruses) will stop dumbfounded at such a strange-looking creature speaking their language. I have watched him carry on an entire conversation with a raft of sea lions, all trying to outbark one another and him. Maldo talks to whales with the same tone and intimations he uses for everyone, because for him, the animal really is a friend. He operates on a higher, spiritual level given to him by a lifetime of magic on the water. His connection to the animals defies logic, and one evening after dinner he told me that perhaps, in a previous life, he might have been a whale. I think he was joking but I believed him. The whales often approach from the rear where the pangero stands, and they will say it is because of the hum of the engine, but I know it is because the whales know these people on a special level.

~

While the lagoon is but a dot on a map, when you stand on its shore it is epic in size. The horizons in all directions except west are outlined with mountains, and from the center of the lagoon, on the water, land is only a thin line in the distance. The lagoon, like much of the Baja coast, is ringed with mangrove forests, called mangals, without which it could not exist. There are more than a hundred different types of mangroves, and each plays an important role in its location. They grow anywhere from six to twenty feet (3–10 meters) tall and have oblong- or oval-shaped leaves. They are salt-tolerant trees known as halophytes, common in tropical latitudes, with a complex filtration system that allows them to thrive in salty conditions a hundred times higher than any other plants could tolerate while also being capable of growing in fresh water. They pass off the excess salt by manifesting it into crystals on the outside of their leaves while taking in the mineral deposits of oxygen generated in their muddy base. Their root system often resembles a map of multiple freeways that serves to keep the trees upright against the violent storms and waves that pound the central Baja coast. While the ones that stand as guardians to the lagoon grow in the thick mud of the littoral, they can also thrive on coral, rocks, or peat.

Mild weather is generated by the cloud cover that arises from moisture given off from the mangrove forests lining the outer shore. Mangroves are easily overlooked because of their constant presence, but they are the true guardians of the lagoon. Their oldest fossils have dated to seventy-five million years ago.

In short, mangroves are an organic, living wall that keeps the coast of the lagoon free from erosion while absorbing the storms and king tides that would overwhelm a less protected area. The center of a mangal is impenetrable to a human. It is a thick, twisting maze of tangled roots whose upper branches provide shelter to countless seabirds, while underwater it is a nursing habitat for sea creatures ranging from gobies to large sharks. Mangroves are highly efficient at storing carbon emissions and play a critical role in stopping climate change.

In recent years, several environmental organizations have partnered to launch the Mangrove Restoration Project in the lagoon. In 2019 they trained a group of local women known as "Mujeres de El Dátil" (Women of the Ratio) on how to collect seeds, use different planting techniques, and maintain mangrove sites. As of this writing in 2022, a documentary is planned by National Geographic to film these women at work in hopes of increasing funding for the planting of some twenty thousand additional mangrove trees.

~

The lagoon has always been a prolific provider of seafood. During the 1980s and 1990s, gill nets proliferated. This, of course, led to the entanglement of not just whales, but other nonedible sea creatures, including dolphins and turtles. Fortunately, gill nets are no longer in use there. The lagoon has long exported their catch worldwide. Four decades ago, most of the bounty would be shipped to the port at Ensenada and then on to Asia, but today, most of it is exported to the United States. A weekly refrigerated truck drives out to the lagoon to load palettes stacked with cases of sea bass, grouper, and lobster. Thanks to whale watching, local poverty has become a thing of the past. The fishermen realized how the nets affected the whales and now they make enough money during whale season that they no longer employ gill nets and can concentrate on fish that bring a higher price, such as sea bass and grouper. While some of them use hooks and lines, the guys at Campo Cortez prefer to use hand lines as their fathers and grandfathers did. I have photos of both Paco and Roberto holding fish the same size as themselves, caught on hand lines.

When the tide recedes, some of us on the camp crew will lead clients on walks across the lagoon floor. Directly in front of our camp, it is totally dry at low tide and we can walk for miles. With each step, angry clams spit up at us as we disrupt their personal time. We dig for fresh scallops the size of silver dollars, and chocolate clams that are pried open and eaten on the spot with salt and a dash of lemon. The scallop shells of the lagoon are shaped like an ice cream cone and can be over a foot long. The thin end is embedded in the sand and only a couple inches of the crown protrudes, so you must know what you are looking for. We use a steel shank with a barbed end that is inserted into the tiny opening at the crown of the scallop. It is then turned sideways to lock the barb inside the shell to pull it up and out of the murky, clinging sand. It can then easily be pried open with a knife. The scallop itself, nestled inside, is the size of the bottom of a coffee cup and up to an inch thick. Oysters are saved to be barbecued later in the evening. If we spot an octopus dwelling, we stick a pinky finger inside until a tiny tentacle wraps around it to be pulled into sunlight, and no, we do not eat one of the smartest critters on Earth, but only enjoy its presence. I can sit for hours and be entertained by gulls dropping clams and oysters on the rocks to crack them open, laughing at incompetent birds who keep dropping the same shell on the sand and wondering why it will not open.

We rarely think of nature as being incompetent but there were two ospreys who gave new meaning to the word. I spent several happy hours one evening watching this intrepid pair who seemed to have missed the genetic codes passed out to the rest of their species. I named them Ricky and Lucy for the 1950s comedy duo for whom nothing ever seemed to go right. My cabin that trip had a skylight, and I could clearly see Lucy's silhouette through it as she patiently waited for her mate to return. Ricky would bring in sizable tree branches, and it was her job to place each one in its perfect spot to interlink the whole into an architectural triumph, but each time Ricky loudly dropped a branch on my roof, Lucy would stare at it while it slid down and off on to the ground. Ricky kept up his routine of shuttling. Each time, Lucy would just stare at the branch as it slid away. With the rhythmic thud of a new branch on my roof I could not have slept if I wanted to. In the morning, there was a sizable pile of tree branches on the ground but nothing on the roof. Ricky and Lucy were nowhere in sight, and I never saw them again. Here's hoping they finally found a home.

~

On one spectacular cloudless morning, we had been on the water for more than an hour without spotting a blow. Maldo decided to head for the *boca*, and since the surf was not as bad as it often is there, we ventured slightly into the open ocean. The first thing I noticed were thousands of cormorants floating everywhere, more than I had ever seen, and I was trying to think of what would attract them when we suddenly had three adult whales next to the boat. Since we were slightly outside the lagoon, they could have been males, because I thought them to be unusually aggressive. The swells had picked up outside the *boca*, and we were tossing around a bit but enjoying the three whales when two more joined them. Suddenly we were being pushed all over the place by each whale who seemed to have its own idea about what to do with us. They began to pass us back and forth, each taking a turn to whirl us around while the sea grew restless. I was losing my balance while trying to video the event and realized we might be in trouble, not from angry whales, but because these forty-ton behemoths were having such a good time playing with us. Whenever I have begun to feel anxious, I look to Maldo. He stands by the tiller like a maritime statue, always smiling, and more often than not, whispering to the whales. Maldo decided the cormorants were feasting on a surface bait ball and that had attracted the whales. But they did not stop there. They followed us back through the surf at the *boca*, bodysurfing as well as any person, and seemingly having a great time. When I looked back at him, Maldo was smiling like a little boy at recess.

Each panga is allotted two hours on the water at a time, usually once in the morning and once in the afternoon, but if we have made good contact, we often stay a little bit longer. Sometimes this has consequences. One day we just sat offshore watching a pair of coyotes digging in the sand. One of them began pulling and tugging and was in the midst of a mighty struggle when suddenly a very large eel popped out of the sand and the surprised coyote fell backward. I estimate the eel was about six feet long and both coyotes took an end of it in their jaws and were engaged in a serious bout of tug-of-war. I photographed the moment, as it was a rare sight, and we were so fascinated we lost track of time. The tide had gone out and we found ourselves stranded in only a few inches of water, not enough to get back to camp. This has happened to me many times over the years and, each time, it's been my job to go over the side and help drag the boat along. I must walk by shuffling my feet because of the numerous stingrays that will strike for sure if I happen to step on one. Their sting is not fatal, but no reason to be careless. I have no idea how many miles I have dragged a panga through shallow water over the years, but each time, it has been worth it. I might add that the same thing has happened to me while kayaking in the mangroves. On two occasions, I stayed too long and had to drag my kayak more than a mile back to camp with the tide out.

~

The lagoon has always attracted its share of colorful characters, some famous, others infamous. One person I would have liked to meet but did not was Christopher Reeve, the late actor of *Superman* fame. Mr. Reeve was an avid scuba diver who carried a particular passion for gray whales. In 1995 he took a journey to produce a documentary for a segment of the popular television show *In the Wild*. Together with a professional cameraman, he began in St. Lawrence Island, Alaska, and followed the gray whale migration route all the way south to San Ignacio Lagoon. Along the way, the footage shows him diving and bodysurfing alongside the grays. Pachico Mayoral appeared with him in the documentary, making tacos together. Reeve was a skilled equestrian, too, who often competed. Shortly before he was scheduled to do the narration for the film, he suffered a terrible fall from his horse that left him paralyzed from the neck down. Fellow actor Stephen Collins did the narration for him. In the documentary, regarding gray whales, he said, "The eye is so big you sense an intelligence in it that there are creatures who have knowledge and understanding that we've only begun to appreciate." Mr. Reeve eventually succumbed to his injury a year before

I started working in the lagoon, but he never stopped being a tireless advocate for the gray whale.

On two occasions, my clients were professional musicians. The first declared on the way out in the panga that she intended to sing to the whales. The entire boat endorsed this but had no idea what was going to happen. We cut our engine to let the tide carry us when this lady, who was not large by any means, stood up in the panga, and with the voice of a Wagnerian Valkyrie, belted out an operatic aria that could be heard across the lagoon. It was like that great scene in *The Shawshank Redemption* when the prisoners all stop to listen to music for the first time over the loudspeaker. I recall two whales spy hopping, just close enough to see what the strange new sound was. Another time, a flautist with the Los Angeles Symphony joined my tour. When we reached deep water and started to drift, she brought out her flute and sent the most delicate sounds skimming over the water like skipping-stone notes. There was a mother and calf not far away and I thought I saw the calf moving back and forth to the music. It was most likely trying and failing to spy hop, but I chose to believe the whale was dancing in its own way.

One young whale that I took to be a yearling stays in my mind. It spy hopped us and began to spin, so I immediately christened it Dervish. Instantly naming whales is always my way to remember them. Dervish got closer and closer to the panga, spinning and twisting in his own improvised dance. There was one elderly gentleman in the rear who was leaning over to try to touch it when the whale accidentally knocked his baseball hat off onto the water. Before we could retrieve it, Dervish had scooped up the hat and carried it around the boat for a minute, showing it off like a trophy before loudly spitting it out. Humpback whales will come up from below their prey and take in tons of water along with sea lions and pelicans unlucky enough to be that close. They have always spit them out right away and I do recall seeing one unfortunate pelican soaring through the air for a good twenty feet before landing in a disheveled heap on the water's surface, covered with whale goo. That day with Dervish was the only time I have ever seen a gray whale spit out anything other than water. The gentleman was elated at the occurrence, saying he could not wait to tell his friends that a whale had tried to eat his hat, and barfed it back up!

I remember two sisters from the San Diego County Sheriff's Office who were obsessed with the idea of kissing a whale. They told me they had come to the lagoon three years in a row but had never had the chance to do so. For them, it had become an annual quest. I happened to be on the water in another boat when the grand moment finally took place, and it was not as elegant as they had hoped. A baby whale popped up right next to their boat. They both tried to touch it, but it stayed just out of reach—baby whales love to tease. As the whale inched closer, one sister lunged out with open arms and went headfirst into the water, bending entirely in half over the boat as her sister hung onto her belt. When she came back up both were laughing hysterically. I never saw either of them get their kiss, so they may still be returning annually in search of their holy grail.

On that same note, there was a fellow from Nebraska who had brought his two pet ferrets with him—that is, if you can consider ferrets to be pets. The man was a mass of cuts and scars while insisting that his ferrets were the gentlest of animals. He kept them in his cabin during the days out on the water, and their ferocious sounds at being locked in captivity echoed across the camp, not to mention their anger upon his daily return. When he left camp, the mattress and sheets were thoroughly slashed from their claws.

Fortunately, we did not allow him to bring the ferrets out in the boat with us.

Then there was the fellow from Pacific Life Insurance Company. He was on one of my fly-in trips and arrived with a cameraman. The company wanted to incorporate a whale that was common to the Pacific Ocean into their logo since they had the name *Pacific* in their title. He told me the company wanted film of breaching whales and he had already filmed breaching orcas and would next photograph humpbacks. We were on the water for three days and saw lots of whales, but for the first two days the only breaches were too far away to be used in a commercial. On the third day, the whale magic kicked in. Not only did two large females breach close by, but one of them did so repeatedly. The gentleman was very excited about his footage and showed it in camp and intimated that it might be possible for all of us to appear in the final commercial. A few weeks later I got in touch, and he told me that the company had run focus groups to get public opinion about which whale should be used. Apparently, the public thought gray whales were "not very good looking," and had "little personality." His words, not mine. I often see the final commercial on television. It begins with an actual breaching humpback whale that, in midair, digitally morphs into the company's logo. I have always thought a gray whale would have done a better job of selling insurance.

One question I am always asked is if I have a favorite whale. After interacting with thousands of them, I can honestly say, yes, I do. I named her Cyclops, and at first appearance she might be considered among the most unfortunate of gray whales. She had only one good eye, the other entirely opaque, which I assume was due to years of scraping the bottom while feeding. Her chassis was a mass of scars in all sizes, a testament to her ability to survive. She had no calf, and that combined with her almost pure white rostrum told me she was elderly. In her prime, she may have produced as many as forty calves, but now she was a lady in decline. I guessed her length to be at least forty-five feet, making her a grand dame of the lagoon, one who had been there and done that, and one could only guess how many times she had made the grueling migration route. Her rostrum was covered with a massive collection of barnacles, almost the size of a beach ball, and her lower jaw stuck out like the jaw of Slackjaw, at an angle that told me it might have been broken and healed incorrectly. She had at least three large shark bites in her left flank that had never properly closed, and part of her right flipper was missing. I was sure she had felt the lancing blade of a whaler because the most telling scar was an angry pink wound the size of a basketball indented into her side. I had seen similar wounds on whales that had been harpooned. She was a reminder that, during her lifetime, gray whales were still routinely hunted. She was ugly even by gray whale standards, but she had heart. That whale had seen the worst and best of humans in her time, and I felt she was telling me her story.

We met quite by surprise one gray morning as she suddenly appeared next to my panga like a silent apparition. I usually see or hear a whale approaching, but not Cyclops. Just unexpectedly, she was there, raising her giant head, not in a spy hop, but more like a world-weary person looking up to see what is happening. It was her stare that captured me. Most whales will look at you then close their eyes. Some of them will keep them open while they spy hop, but no whale had ever maintained the continuous eye contact I had with Cyclops. I was convinced she wanted to communicate, and I cannot describe the sensation that pulsed through me when our eyes met. I will risk using the word *telepathy*; it was that strong.

My mind was spinning after that initial encounter, and I was sure I had done some head tricks on myself as I felt her standing in silent judgment of me and issuing a blanket forgiveness. In my business I have found it impossible not to occasionally impose my thoughts on an animal, hoping them to be true. I saw her each day for four days and her behavior never faltered. She would approach, raise her head, and stare at me, willing me to understand what she would have me know. I was filled with empathy, and I tried to imagine a long lance entering my side, the searing pain like fire as the razor-sharp barb lodged inside my body, while everything around me faded into black and my life force drained out into the ocean. In that short time together, I felt I had absorbed the entire spectrum of whale suffering and elation, all through the eyes of this matriarch. Logic tells me this was all a figment of my imagination, my wishes for a daydream, but no other animal anywhere has ever had such an effect on me.

On my last day of that particular trip, she seemed to sense that our time together was over. Instead of her usual staring, she rose mightily from the water and gently leaned over, resting her massive head on the gunnels of the panga, tipping us over precariously, right to the water's edge, but not enough to dump us. I leaned in, arms spread wide, and hugged her as best I could, our eyes resting inches from each other. She seemed to search my eyes in one final hope that I knew what she was saying. Today, years later, although rarely, Cyclops still enters my dreams, and I can sense myself on the water with her. If you believe something strong enough, it can become real.

~

When Irene and I returned from our first trip to the lagoon, we talked about it so much that two of our pals said they wanted to join us the following year. We took that with a grain of salt, because our friends were not what one would call outdoor types. They were more Rodeo Drive types. When the time rolled around, we were surprised to find they were still requesting to go with us. At that point we had to be diplomatic. We were sure they would not be comfortable in the rustic camp that at the time was still using the old canvas surplus tents. We tried to describe the absolute wilderness we were going into and total lack of the types of amenities they enjoyed. Still, they asked to come; so, we said okay. We were still new to Baja ourselves and there were always rumors of kidnappings and robberies, and although we never experienced anything of the kind, we still took precautions such as asking our friends to wear neutral-colored clothing and refrain from wearing jewelry so as not to draw unwanted attention. To that our lady friend replied, and this is a direct quote that I will never forget, "I am only going to bring my hiking pearls!" And she was serious.

The five of us—Irene, me, Barbara, Dave, and Peter (not their real names)—all flew from Los Angeles to Loreto where Johnny Friday was waiting this time and piled us into the dilapidated van for the six-hour ride to the lagoon. There is darkness and then there is Baja at night, especially traveling through the desert on undeveloped roads. We could not see a foot in front of us and hoped we did not run over any critters. It was a real bone shaker, which Irene and I were used to but our friends were not. We arrived in the middle of the night to a howling and frigid wind with our friends wondering what they had gotten themselves into. At the time the only lights were kerosene lanterns, and the only heat was from layers of fleece. We huddled in our respective tents as the wind screamed like banshees, flapping the canvas like crumpling tinfoil. I remember Dave telling Barbara that in the morning he would do whatever was necessary to get them out of there. Peter said nothing at the time, but I expected him to go with his friends if they left. It was the worst-case scenario of getting your wish and being sorry for it. After a mostly sleepless night, the Lord took pity and gave us a beautiful morning. Things looked so good in fact that we found Barbara putting on her makeup at the outdoor wash basin. Until we mentioned it, she was not aware that a half dozen fishermen were ogling her, having never seen a woman in camp with all that makeup before. I think she left the hiking pearls in the tent.

Next, the five of us put in with Maldo's panga and headed out for a hopeful day. Our friends seemed uneasy but were keeping up good spirits under the circumstances. When a whale approached, everyone kept their seats while Irene and I made fools of ourselves petting and baby talking a forty-ton animal. After a short while, Dave decided to go for broke and cautiously leaned over the side where a yearling had been hanging for several minutes, mouth open, to have his tongue scratched. Dave slowly sent out a most tentative reach with pointed finger he may have thought would be bitten off, and after the briefest of contacts, he pulled back as if he had touched a hot stove. He stood there in the middle of the boat as an enormous smile slowly crossed his face and he said in a tiny voice, "I touched it!" Once was enough for him, but Dave spent the rest of the morning smiling.

That felt like a coup, but the day was only beginning. We had been out more than an hour when Peter announced he had to use the bathroom. We were not going to motor all the way back and lose our allotted time on the water, so Maldo told us all to look the other way while Peter stood at the rear of the boat to relieve himself. Just to be safe, Maldo held him by his belt so he could lean over the stern, and at that exact moment, a young whale shot out in a lunge and hung there taking a shower. That is not a story I could make up. I have too many witnesses. That combined story has fueled countless dinner conversations over the years, and today, Dave, Barbara, and Peter are still our good friends.

NINE

Protecting the Helpless

In 1946, with the world still emerging from a global war, some people began to appreciate that whale populations had increased during the war years because whaling had ceased. In the first true effort to protect these growing stocks of animals, the International Convention for the Regulation of Whaling (ICRW) was held and resulted in the formation of the International Whaling Commission (IWC). Its charter stated that its purpose was to "Provide for the conservation of whale stocks and thus make possible the orderly development of the whaling industry." The IWC is still active today, seventy-five years later, based in Impington, United Kingdom. It was meant to be a blanket organization that governed the worldwide taking of whales, granting protection to specific species while setting aside designated areas as sanctuaries. It was supposed to declare when and which kinds of whales could be hunted, and how to keep active whaling sustainable. Most important, it prohibited the taking of calves or mothers with calves. It took on whale research and the publication of the results and even introduced studies about the cruelty of whale hunting. Over the years, the IWC evolved to include research into net entanglements, ship strikes, the effects of noise, air, and water pollution, how to safely conduct sustainable whale-watching operations without harassing the animals, and myriad lesser issues. Membership was not initially limited to nations that were actively whaling and has continued to grow steadily as people have become increasingly educated about these intelligent animals. As of today, there are eighty-eight member nations.

For the past decade, the IWC has met every other year, with each participating nation sending one representative with an unspecified number of "experts" to argue their case for or against their chosen cause. As in any massive bureaucracy, most of these meetings have been quite divisive and often result in a member walking out. There seems to be an almost even divide between nations that wish to continue commercial whaling and those that wish to stop it as the theory of whales being sentient beings has spread rapidly in just the past year. A 2018 meeting of the IWC produced a document known as the Florianópolis Declaration, which gave *protection to all species* and allowed for populations to increase to the numbers reached before industrial whaling. Three separate types of whaling are recognized by this convention: commercial, aboriginal subsistence, and special permit (more commonly known as "scientific"). It is the latter that has allowed Japan to sidestep the IWC regulations. As a result of this, Japan created an exclusive "economic zone" in its home waters and announced that it would resume whaling operations. The Japanese appetite for whale meat has long been among the most voracious among whaling countries. Friends of mine who were inspectors for the IWC bought random cans of whale meat in Japanese markets and found it not only contained meat from multiple whale species, but also that of animals on the endangered species list. This must be viewed in a cultural context.

Island nations such as Japan have traditionally been great consumers of seafood, and it is difficult to change thousands of years of culture in a few lifetimes. Norway, Iceland, and Tonga, among others, are culturally traditional whaling nations as well.

It was only in 1986 that the IWC called for a ban on all commercial whaling, and while most countries complied, many did not. Whale populations increased, but never reached their preindustrial numbers. The problem with all of this is the lack of a governing body that can enforce the regulations. Each signatory member of the IWC pledges to follow its rules, but there is no policing body to make sure they do it.

~

The Marine Mammal Protection Act is a separate declaration from the IWC and became federal law in 1972 during the Nixon administration. Its purpose is to maintain the health of a stable marine ecosystem within the waters of the United States and its territories. Under its umbrella, it is illegal to "harass, feed, hunt, capture, or kill any marine mammal." This includes cetaceans (whales, dolphins, and porpoises), sirenians (manatees and dugongs), and pinnipeds (seals, sea lions, fur seals, elephant seals, and walrus). It also covers sea otters and polar bears within United States territorial waters. Mammal rescue centers are granted special permission to respond when marine animals are stranded. Before the MMPA, marine mammals were managed according to the value they were thought to provide to humans, such as performing in aquatic theme parks. The MMPA designated marine mammals as wildlife rather than property for the first time. The organizational bodies associated with the act form such a large bureaucracy that one can get lost navigating who oversees what.

There have always been exemptions to the IWC and MMPA rules. Countries with histories of subsistence whaling that claim it is integral to both their diet and culture have always been allowed to take a certain number of whales. These include a number of island nations such as Tonga, the Inuit of the Pacific Northwest, several Canadian tribes, and the Makah tribe of Washington State.

≈

Five tribes make up the Makah nation, which has occupied the northwestern corner of the United States for thousands of years. Their traditional homeland is near Cape Flattery in Washington State, part of the Olympic Peninsula where the Pacific Ocean collides with the Strait of Juan de Fuca. In 2007, after almost seventy years of a self-imposed ban on whaling, the tribe decided to kill a whale once again. It was to be a gray whale.

Through much of their history, the Makah approached whaling as a spiritual endeavor heavily layered with prayer and ceremony. A typical hunt would begin several days before the actual killing, with the whalers bathing in the frigid northern waters and using shells to scrape their skin raw for purification. The most intense among the tribe invoked their ancestors to journey with them onto the water for the hunt. Each whaler answered his own inner call, going off into the forest to pray and bathe for several days. They went to sea in traditional hand-carved cedar canoes up to thirty feet long paddled by eight men. Harpoons could be up to six meters long, made from a hard yew wood with barbs on either side of the tip made of razor-sharp mussel shells. The shaft was spliced so that it would break away if the whale became violent and threaten to hit the boat with the shaft. Once it had penetrated the whale, additional barbs made from elk antlers anchored it in place. The hunters would cover the harpoon tip with pitch from a spruce tree so it would enter the whale's flesh easily. The killing end was attached with whale sinew and sometimes wrapped with bark to hold it in place. As much as 250 feet of coiled rope, made from twisted cedar boughs, would be stored in a woven basket that played out smoothly as the whale ran so as not to entangle any member of the crew.

When they found a gray whale, the boat master would approach from its left side, gauge its speed, and then direct the harpooner in the bow to sink the shaft into the animal. A float made from a seal skin would be attached to the end of the rope to show where the whale was and to provide drag when it tried to dive, causing it to tire rapidly while losing blood. Once the whale expired, a crew member would enter the water and stitch the whale's mouth shut to prevent the sea from entering and sinking the whale. Prayers would be offered for the spirit of the animal. After a successful hunt, the crew would often sing a traditional song in unison to celebrate and to notify the tribe of their return.

When Europeans began arriving in the seventeenth century, the Makah actively traded whale meat and oil with them through the following century. In 1855 the tribe entered into a treaty with the territorial governor of Washington, which was not yet a state. His name was Isaac Stevens, and the agreement was the Treaty of Neah Bay. The tribe agreed to give up much of its land in exchange for the right to continue taking whales. That treaty remains to this day the only one of its kind allowing a native American tribe the right to take whales. Its only proviso was that they could not trade the whale meat on international markets, which they had no intention of ever doing.

Even though they now had the legal right to take whales, the tribe abandoned whaling for the next three decades, focusing on taking fur seals instead. This source of food, while lucrative at first, was depleted by the end of the nineteenth century and then outlawed by the US government, forcing the Makah to return to whaling, but only on a sporadic basis. By this time, commercial whaling had depleted most of the local gray whale stocks. The Makah only occasionally killed a whale until the early part of the twentieth century. They fell on tough times. So they began to think about reverting to their whaling heritage, but had no idea how public opinion had shifted over the years.

On May 10, 1999, the Makah set out in a traditional cedar bark canoe named *Hummingbird*, looking to kill a gray whale off the Olympic Peninsula, and unleashed a firestorm of public outrage. More than 350 opposition groups from twenty-seven countries publicly castigated the Makah, with the loudest voice being that of Paul Watson, the founder of Sea Shepherd, one of the more militant of the anti-whaling organizations. The protesters argued that US government support of the Makah would jeopardize their efforts to end commercial whaling and feared it might incite dozens of indigenous tribes to also begin taking whales. Dozens of small craft operated by the protesters assigned themselves to shadow the Makah canoe, making as much noise as humanly possible. The US Coast Guard was on hand to oversee the situation. They were joined by Sea Shepherd's massive ship, *Sirenian*, along with a most unusual protest tool—an aging submarine painted to resemble an orca with loudspeakers on its deck that blared orca cries at ear-splitting decibels, hoping to scare any potential whales away. Every time the *Hummingbird* left the dock, the submarine would shadow it until the Coast Guard made them stop, citing them for harassing the whale that the Makah were trying to kill!

Since the Makah had gone out twice without making a kill, the *Sirenian* left Neah Bay for business reasons. During its absence, the *Hummingbird* managed to slip out in the early morning fog of May 17, eluding its entourage of protest boats. An overhead helicopter recorded events as Makah whaler Theron Parker sunk a harpoon deep into the side of a thirty-foot gray whale. The harpoon did not shatter as planned and Parker grabbed it, pulled it out, and plunged it in a second time at which point a Makah rifleman in the *Hummingbird* issued the coup de grâce in the form of a .50-caliber bullet from a high-powered rifle. So much for tradition and ceremony.

Grays were hunted again later that year, and the whalers were always confronted on the water by protesters. That summer, a court of appeals sided with the protesters and suspended all whale hunting by the Makah. In 2007, after a lengthy legal process of appeals, several tribal members ran out of patience. On September 8, five young Makah killed a whale in the Strait of Juan de Fuca. The whale had been impaled by a harpoon as well as shot several times with a .50-caliber rifle. It sank while still alive. There was nothing spiritual involved with this kill; there was no ceremony or purification, no fasting or prayer. It was as Makah elder Roberta Thompson told me years ago in Neah Bay: "Some young men . . . wanted to go out and kill something."

Neither the National Oceanic and Atmospheric Administration (NOAA) nor the Makah tribal council had authorized this hunt, since it violated several restrictions from the original whaling application submitted by the tribe. In October, the five men were indicted in federal court on charges of conspiracy to engage in illegal whaling, unauthorized taking of a marine mammal, and unauthorized whaling. Their own tribal court charged them with reckless endangerment and violating the tribe's gray whale management plan, along with several state and local laws. At trial, three of the whalers pleaded guilty to violating the MMPA by the unlawful taking of a marine mammal when, in return, prosecutors agreed not to ask for jail time. At a separate bench trial, the remaining two who did not plead guilty up front were convicted of the same charges along with conspiracy. The first three were sentenced to two years of supervised release plus various hours of community service. As for the remaining two, one got a sentence of five months in prison, a year of supervised release, and one hundred seventy-five community service hours. The other got ninety days in jail and two hundred community service hours. The presiding judge also forbade any of them from taking part in any future whale hunts, legal or not.

The whale killed by the Makah in 1999 was the first and only one successfully taken by the nation since 1920. The bones of that whale reside today in the Makah Cultural and Research Center in Neah Bay where local high school students meticulously reassembled them. A Makah elder who asked to remain unidentified put it well: "We have not killed whales for over a generation. There is no need to start again now."

~

Gray whales have, for the most part, avoided captivity for several reasons. First, they are not as beautiful looking as the dynamically colored orca or the snow-white beluga. They are not as intelligent as either species, so they cannot be made to perform tricks, and they are much larger than either of the other whales, requiring a larger enclosure and additional food consumption. Orcas will perform like trained dogs to receive a food reward, and belugas, with their permanent smile, are often seen looking back at the people watching them, interacting with them. Gray and other large whales simply do not have the charisma of orcas or belugas, and as a result, they have been, for the most part, spared incarceration. But there are two glaring exceptions.

Aside from the cardiac studies conducted in 1956–1957 by Dr. Paul White, which produced limited results, almost all of our physiological knowledge of gray whales comes from two young captive grays named Gigi and J. J.—taken from the sea decades apart. In 1971, Gigi and her mother were in the nursery lagoon of Ojo de Liebre, the lagoon north of San Ignacio in Baja. Gigi was forcibly taken from her mother by a ship's crew from San Diego's SeaWorld Park for "scientific study." A personal friend of mine who was on the capture team told me the mother whale beat her tail flukes against the capture ship for hours in a futile attempt to rescue her calf. The captors likely thought of these creatures as dumb animals and had no empathy for splitting up a real family.

Gigi was only six weeks old, a bit shy of the usual seven to eight months it takes to ween a gray whale calf. Gigi was the largest marine mammal to ever be kept in captivity and no one had a manual on how to care for a baby gray, especially one not even old enough to act like a whale. The SeaWorld crew had to work around the clock to keep her alive. What helped was spreading fish parts over the bottom of her holding tank. She had spent enough time with her mother to know how to feed from the bottom and this may have saved her life. As she regained her health, she was put into a tank to be ogled by crowds that at the time had no idea how this animal was being tortured. The space she occupied was not much larger than a normal residential swimming pool. She barely had room to move a few feet in any direction. Fortunately, she survived to be released successfully into the wild one year later. It never occurred to her keepers that migrating animals need to migrate. If anything good came out of that incarceration, it was increased knowledge of the physiology of Gigi's kind.

In 1997, a starving and dehydrated young gray whale was found washing back and forth in the surf line off the shores of Marina del Rey, in Southern California. Too young to be on her own, she was helpless in the crashing surf, and mother was nowhere in sight. If she had not been spotted, she would have drowned. Old video footage shows local people coming to her aid, trying to steady and calm her until a marine mammal rescue team arrived. Local people covered her with blankets kept wet with seawater and made sure nothing obstructed her blowhole. By the time SeaWorld got a van to lift the pathetic critter, she was comatose and close to death. They called this whale J. J., and after rehydrating her, they used a tube to force-feed her. Surprisingly, she gained about fifty pounds a day on a diet of condensed milk and fish parts, about half of what she would gain in the wild, but she survived. Fourteen months later, J. J. weighed in at about nineteen thousand pounds and was much too large to keep in her pen. She was at least twice the size of any captive orca. A tracking device was attached to her back, and she lost it within a day. Her "experts" did not realize how much gray whales like to rub against things. She was released near a pod of gray whales and was never seen again. Whether she joined the pod or not will never be known. She could have been photographed to document her markings, coloration, scarring, birthmarks, and so on, but biologists were not doing such science in those days; if they had, she might have been located over the years.

≈

The United States last issued a permit for the taking of a wild orca from the sea in 1989. Other nations not bound by American law continue to capture them for theme parks, especially China where such parks are proliferating. Again, these are regulations without any teeth. There is no central authority, like the Navy or Coast Guard, on the water making sure these violations do not occur. It is almost entirely enforced by individual whale-watching operators. Both the American Cetacean Society and the whale camps of San Ignacio observe these regulations vigorously, as have most American whale-watching operators. Still, the MMPA is a study in contradictions. While it is illegal to take a wild orca in American waters, it is not illegal to display them. On top of that, the laws governing the housing of these animals are woefully inadequate.

The first captive whale was a female orca nicknamed Wanda, taken on November 18, 1961, in the cove below the now defunct Marineland of the Pacific in Palos Verdes, California. A huge crowd gathered onshore to cheer the whale as it put up a valiant fight, repeatedly evading the capture nets and lasso nooses for an entire afternoon before surrendering in exhaustion. Just before being hauled onto the capture vessel, several crew members observed her upside down, repeatedly striking the water with her tail flukes, a definite sign of distress.

She was hauled to the marine "park" and put into an oval tank measuring 100 x 50 x 19 feet, and immediately began swimming in circles, continuously ramming into the walls in her attempt to escape. Suddenly, this highly social creature, used to roaming hundreds of miles in a day, was confined to a tiny pool, in complete solitude, except for people who wished to make her perform tricks as entertainment. Only two days later, Wanda began circling her pool at high speed before ramming the wall with her head, after which she convulsed and died. As of this writing, there are believed to be approximately thirty-six hundred captive whales and dolphins worldwide, and as of 2021, fifty-seven of them were orcas, with thirty born in captivity. Also as of this writing, SeaWorld Parks are believed to hold nineteen of them. Those whales will never know the open ocean and will never truly know how to act or feel like whales.

How many reside in China is unknown. It is believed that some are held in the aquatic "parks" constructed in recent years there, as they would require captive whales in order to turn a profit. Evidence points to Russia as the illegal supplier of captive whales, but few have had the nerve to speak publicly about it until the film I wrote about earlier telling the sad tale of illegal Russian whaling. Whales come even lower on the pecking order when Russia is busy conducting an illegal war in Ukraine.

There is only one reason to put an animal used to swimming hundreds of miles a day into a cage, and that is money. Keeping an orca in such conditions is no different than locking you or me inside a closet 24/7. It is simply torture. To take away the natural instincts of any wild animal is to reduce it to human dependency, and that usually leads to cruelty.

Current laws allow for the capture of whales for scientific research and "entertainment." It is also not illegal to make them perform tricks, but there are laws governing their housing, and yet, there is no tank on this planet large enough to house an orca in comfort. In the confines of an aquatic park, it is common for the so-called trainers to stand on an animal's back or have it lift them out of the water by balancing on the end of its rostrum. Orca and dolphin skin is unusually sensitive, and this often results in open sores. All so the paying public can applaud its complicity.

The first sign of a sick orca is a drooping dorsal fin, sometimes followed by a rash on the base of the whale's saddle patch. The dorsal of an adult bull orca can be up to six feet tall and biologists believe that, in the wild, the whale's constant movement helps to keep it upright and rigid. Confinement in a small pool just does not give the same hydrodynamic resistance, and the dorsal folds over like a wilting flower without sun. If you look at video of any orca in captivity, it will have a drooping dorsal (meanwhile, all of them have mental issues). The best-known example of this is in the movie *Free Willy*. Even the movie poster shows a drooping dorsal, and illustrates the indifference of people, especially the so-called whale experts of that time who should have known this, or just did not care about the suffering of an animal.

The whale that stars in *Free Willy* was a two-year-old orca taken from Icelandic waters in 1979. It was assigned the name Keiko. During the next six years, Keiko went from one tiny tank to another between Canada and Iceland and was eventually sold to a Mexican amusement park. The whale was made to perform tricks in a tiny pool from 1985 to 1996, eleven more years of solitary isolation. By this time, the animal did not know how to act like a whale; it could only do what it was taught.

In 1993, Keiko was chosen for *Free Willy* and was obviously in distress for the entire movie. The movie was a box-office hit, however, and generated such a public outcry for his release that, in 1996, Warner Bros. collaborated with the International Marine Mammal Project to return him to his home waters. He was sent to a rehabilitation facility at the Oregon Coast Aquarium in hopes of getting him fit enough to return to the wild. For three years they trained him to follow their boats, hoping to get him used to swimming in the open ocean, but once finally released, he was only observed swimming with other orcas one time before disappearing for weeks.

He ended his days in the open ocean of Norway in 2003, alone, not knowing how to be or act like a whale. He probably died of starvation, since a later necropsy showed almost no food in his stomach. If anything, it was one of the first times that truly serious attention was brought to the plight of captive whales, and it hopefully contributed to the public's enlightenment of these animals as fellow sentient beings.

In 2010, a malnourished and underweight orca was captured off the coast of the Netherlands in the Wadden Sea and assigned the name Morgan. The whale was "owned" by SeaWorld but was "loaned" to a Spanish aquatic park, Loro Parque, in Tenerife, in 2016 to participate in its breeding program with the proviso that she would not be put on display or made to perform tricks, and that she would eventually be returned to the wild. According to the Dolphin Project, Morgan immediately began ramming her head into the sides of her holding tank and showed signs of anxiety and stress. When Morgan could take no more, she "beached" herself on the side of the pool and lay there waiting to die. A beached orca will suffocate from its own weight when out of the water. The Dolphin Project released video of this attempted whale suicide. Loro Parque director Wolfgang Rades issued the ridiculous statement that it is normal for orcas to beach themselves, especially when hunting. He also added that Morgan was not unhappy because there was much more to see when outside of her holding tank.

In 2017, SeaWorld ended its captive breeding program, and ownership of Morgan was officially transferred to Loro Parque. In 2018, Morgan gave birth to a daughter named Ula that died recently. Morgan continues to suffer in confinement and is not the only whale to try to take its own life while in captivity. Two separate orcas, Hugo and Kiska, both committed suicide by ramming their heads against their tank walls until they had brain aneurysms and died. The *US Sun* released a study in 2017 that claimed 25 percent of captive orcas had ground their teeth on the sides of their tanks, and many were given Valium to try to calm them. Marine scientists have equated such behavior to post-traumatic stress in humans.

Xplore Our Planet, a resource site for wildlife enthusiasts, constantly updates information about captive whales. In January 2013, the documentary film *Blackfish* premiered, telling the story of a captive orca named Tilikum that rebelled against his treatment and was involved in the deaths of one of his "trainers," along with a park trespasser. Almost overnight after the film's release, aquatic park attendance plummeted. SeaWorld agreed to end its captive breeding program, and the public began to understand that keeping whales captive was inhumane. *Blackfish*, more than anything, opened people's eyes about intelligent creatures in captivity.

Fortunately, according to Xplore Our Planet, cetacean captures, trade, and sale are now illegal in most countries. More important, breeding orcas has halted in the United States. Along those same lines, Iceland, too, as of this writing, has agreed to finally end commercial whaling in 2024.

Today, should the world awaken and realize what has been done to these animals, it is still too late for those currently living in tanks. They will never return to their pods or even the open ocean again because living in such limited isolation destroys natural instinct, especially hunting techniques. The whales currently in captivity will all die there.

~

Since dolphins are the smallest of the toothed whales, I must include them here, especially since breeding captive dolphins has remained a popular practice for parks that allow people to swim in tanks with them, and public attendance at such parks is rapidly increasing. In the 1960s, Ric O'Barry spent time working at the US Navy Marine Mammal Program in Point Loma, San Diego, California. This organization trains dolphins and sea lions for use by the Navy and is overseen by the Naval Information Warfare Center. Mr. O'Barry also helped to capture and train dolphins for the Miami Seaquarium. Five dolphins were trained to play the part of Flipper in a popular television show of that era. Each of these dolphins played the star at different times, since, being food driven, once fed, that particular dolphin would no longer perform.

One day, the main dolphin used for filming, named Kathy, submerged and refused to surface to breathe, effectively committing suicide by drowning, the first recorded such act by a cetacean. Kathy died while O'Barry was holding her, and he then had an epiphany. Realizing the cruelty of keeping wide-ranging animals in confinement led him to found the Dolphin Project, an organization dedicated to the preservation and protection of these mammals, and today it is more active than ever in promoting public awareness of these fragile creatures.

It has long been public knowledge that, beginning in 1960, the US Navy began in-depth research using dolphins. These dolphins were officially called "Advanced Biological Weapons Systems." The first known "draftee" was a bottlenose dolphin named Tuffy, who was used to ferry information and tools to divers on the seafloor in the Navy's Sealab project.

The Navy's own Information Warfare Center (NIWC) has an entire website about the animals they use and the jobs they make them do. They compare captive marine mammals to service animals, such as dogs. That by itself is not necessarily a bad thing as it is common knowledge that intelligent animals are happiest when they have a job to perform. When the project began, they tested a strange assortment of sea creatures for their sensory and physical capabilities—sharks, rays, turtles, and even seabirds. They researched theories about animals navigating by using the magnetic fields of the Earth while watching to see what creatures might be compatible to work alongside humans in the ocean. Today they use only two species of animals, the bottlenose dolphin (*Tursiops trancatus*) and sea lions (*Zalophus californianus*). Why do they use these animals? Dolphins have the most sophisticated sonar in existence, and in fact naval sonar was developed by studying how dolphins echolocate (bounce an electronic impulse off an object to send back a mental picture to the brain). They see easily in dark water and have a keen sense of directional hearing. Both species can dive hundreds of feet without getting the "bends" like a conventional diver. That makes them perfect for underwater work.

The Navy says dolphins locate and mark the location of underwater mines and are trained to ferry tools and oxygen bottles to divers, while also being used as patrol animals for security around ships, much as the Army and Marines use sentry dogs for security. I have also heard stories of the Navy training dolphins to attach explosive devices to enemy vessels but have found no proof whatsoever of that. When needed at a location, the dolphins are transported inside fiberglass containers filled with water to support them in fleece-lined hammocks. They are also accompanied by a veterinarian and a trainer with whom they are familiar. This is all humane enough.

Since 1973, the Navy has had to contend with rumors of cruelty, originally sparked by a popular movie, *Day of the Dolphin*, in which captive dolphins were confined in an exceedingly small pool, studied, and taught to speak.

Since the entire naval project was classified, much of the available information on the internet may be outdated, but in 2017 the International Mammal Project claimed the Navy held eighty-five dolphins in tiny pens at the San Diego facility. The Navy had also contracted with SeaWorld for a breeding agreement to raise confinement-bred dolphins for warfare. Yes, it's the same SeaWorld covered in the film *Blackfish*. In April 2017, a forty-six-year-old dolphin named Makai was euthanized by the Navy for "old age," but the story opened a can of worms as protesters and activists descended on the naval facility.

This public outcry persuaded the Navy to drop the title of "biological weapons" and to rename the animals Mark 1, Mark 2, and so on, hoping the public would not get too attached to any of them if 1) it didn't sound as if the animals went about blowing things up, and 2) if they had no unique names.

While on the open ocean, the Navy fits each dolphin with what it calls an "anti-foraging" device, which is a simple strip of orange Velcro wrapped around the rostrum to prevent the dolphin from eating. If one gets too far away, the Navy uses something called a "recall pinger" that the dolphin can hear at great distances. If the dolphin returns, it is rewarded with food. If not, it is simply replaced with another dolphin.

Because most of the world now knows how the Navy uses these animals, wild dolphins are in mortal danger any time they approach a ship on the open ocean. If it is a ship from a hostile country, the crew may simply blast any approaching dolphin out of the water.

Granted, the Navy has learned much from their captives, especially in the field of sonar where they attach microphones to their jaws with suction cups to monitor their echolocations, but this brings up the question of why animals must be weaponized. Certainly, technology is at the point where underwater drones or similar robotics could be used for much of this work.

I don't believe the Navy is deliberately cruel, but I do fault it for its inability to realize that no pen is large enough to comfortably house a dolphin. These animals range far and wide in a day and need open space. The frightening thing is that aquatic theme parks in the United States may be the most humane versions of these attractions in the world.

Since aquatic theme parks do not always record births or deaths, it is impossible to know for sure how many dolphins are being kept worldwide, but according to IDA (In Defense of Animals) there may be up to three thousand with approximately five hundred being held in the United States. IDA has also listed symptoms of depression commonly observed among captive dolphins. These include logging on the surface rather than swimming, chewing on any protruding object in their pen, constantly swimming in the same pattern (in a tank, are there really any other options?), and head bobbing or jaw popping.

The target audience of these parks is young people, generally small children who are innocently thrilled to see large animals performing. Exposing the very young to captive animals overtly encourages them to form an emotional bond. That is why you see children leaving the park with plush toy animals of the kind they saw performing. They identify with these animals and of course are still too young to understand the morality of keeping animals confined. Such exposure equates love for the animal with human domination, and makes it easy to normalize the idea that some animals should be dominated and confined.

~

"Sentient beings" is a relatively new term that has become a fixture in the general lexicon. It has been applied to animals of obvious intelligence, and interpreted in various ways. In 2017, the Sentient Institute was founded as a vehicle for acknowledging animals as creatures capable of having both positive and negative experiences, such as suffering and happiness. This is the simplest definition of the term. Sentience is closely allied to the philosophical definition of consciousness, which involves a thought process, although it is a more specific term relating to an animal being able to feel pain physically *and* mentally. Few Eastern bloc countries share the growing Western view of cetaceans as sentient beings. As recently as February 2022, the BBC reported that at the world's largest science conference, an annual meeting of the American Association for the Advancement of Science (AAAS) in Vancouver, Canada, several leading conservationists, philosophers, and animal behavioral scientists voted for a Declaration of Rights for Cetaceans. They claimed that long-term research has proven that most dolphins and some whales have a sufficiently complex brain that they can assume a humanlike level of self-awareness. Research has shown dolphins to have the ability to understand concepts such as the continuity of numbers and maybe even the recognition of individual numbers, while not actually being able to count. Most dolphin researchers rank their intelligence on a par with elephants, once thought to be the smartest creatures of the animal world.

Tom White, an ethics professor from Loyola Marymount University in Los Angeles and author of *In Defense of Dolphins*, has stated that dolphins are "nonhuman persons." Professor White also contends that "If individuals count, then the deliberate killing of individuals of this sort is ethically the equivalent of deliberately killing a human being."

As far back as 2010, the AAAS agreed to the following statement: "Every individual cetacean has the right to life, no cetacean should be held in captivity or servitude, be subject to cruel treatment, or be removed from their natural environment." They also added, "No cetacean is the property of any state, corporate human group, or individual." AAAS whale advocates promote the rule that individual self-awareness is "no longer a unique human property." One can hear in this statement the influence of French philosopher René Descartes's maxim, "I think, therefore I am." Studies have proven conclusively that dolphins not only recognize themselves in mirrors, but employ actual names for one another.

Just like human prisoners, cetaceans become institutionalized over time, requiring a rigid daily routine in order to function. Without being told what to do, they no longer have a clue. Most cetaceans now in captivity have been there for years. Their natural instinct has been dulled to the point that they no longer act as the rest of their species does in the wild. They lose the ability to hunt and become totally dependent on humans for food and care.

Since the *Blackfish* movement, the concept of cetaceans as sentient beings has spread. In 1997, the European Union was the first large-scale organization to recognize sentient status for all animals. In 2015, New Zealand, France, and Quebec, Canada, amended current laws to recognize all animals as sentient beings. In 2018, Slovakia joined their ranks. As of 2019, thirty-two countries have formally recognized animals as nonhuman sentient beings. Currently there is a proposal in the United Nations to adopt the "Universal Declaration on Animal Welfare," which would recognize all animals the world over as sentient beings. If it is endorsed, it would set forth a list of principles to acknowledge the importance of animals and the responsibilities of humankind as their stewards. Such enlightened thinking seems to be the first true light at the end of the tunnel.

TEN

Threats to the Lagoon

According to the Environmental Protection Agency, from 1901 to 2020, ocean temperatures increased by 0.14 degrees Fahrenheit each decade. While carbon emissions are the main culprit, other accompanying threats include acid rain, sewage flow from large cities directly into the ocean, oil spills, mercury poisoning from the use of coal, and the creation of dead zones at river mouths from the use of fertilizers.

Krill are tiny shrimplike crustaceans that consume plankton. They are extremely sensitive to temperature change and pollution, and the current temperature rise is rapidly shrinking both their territory and food source in the Bering and Chukchi Seas while they are at the same time dying from the above-mentioned pollutions. The diminishment of phytoplankton, which happens to account for almost half of photosynthesis of Earth, means the young larvae of krill and other amphipods have less and less food. Krill is also the main part of a gray whale diet, which consists of 90 percent amphipods, according to the San Ignacio Laguna Group, and that means the grays' summer feeding grounds are shrinking accordingly.

Such is the cause of an unusual die-off of gray whales recently.

Increasingly, grays have been found emaciated on beaches. The Laguna San Ignacio Ecosystem Science Program has been tracking grays for almost forty-five years, and in 2018, using photo identification, they noticed a 25 percent increase in emaciated whales along with a significant reduction in cow/calf pairs. Scientist Steven Schwarz of that group has written that, "In large mammals, if you don't have the health or energy to bring offspring to term, you'll abort it and preserve yourself."

~

The year 2017 may be remembered as the "still not too late" point to halt climate change. That year, arctic ice covered only about 770,000 square miles (1,994,300 square kilometers). In the Chukchi Sea, there is an area known as Hope Canyon that, thanks to the convergence of currents, has always been a favorite summer feeding ground of the gray whales, but scientists have noticed the whales going farther out from it, a sure sign of food shortage.

In 2019, the term *unusual mortality event*, or UME, became the buzzword. Throughout recorded history there have been periodic die-offs of gray whales that have for the most part been considered nature's way of culling the herd, but that year, 2019, set a new standard. Since 2019 there have been at least five hundred stranded gray whales, and there is no way of knowing how many others sank at sea without being counted.

And this problem is not relegated to the whale's summer feeding grounds or its migratory path. The San Ignacio Laguna Group began finding more and more emaciated dead whales in the lagoon itself.

Grays usually do not feed during their migration, but over the past three years, emaciated whales have been observed leaving their traditional route in search of food. Being opportunistic feeders, they have been coming into bays and marinas. In 2019, no fewer than fourteen whales washed ashore along the coast; six were emaciated. Fast-forward to 2022. A gray whale was spotted inside San Francisco Bay several times over a forty-day period. It may have just been lost but was most probably in search of food. When it finally washed ashore near Tiburon, it was half of its normal size.

In the midst of so much bad news, there is at least one small pod of gray whales that has taken dwindling food supply matters into their own flippers. Some three decades ago, about a dozen whales began deviating from the normal northern migration path and started entering Washington State's Puget Sound near Camano and Whidbey Islands. Locals have nicknamed them the "Sounders."

They come into tidal mud flats, often in no more than five or six feet of water, not even enough to cover their bodies. They turn on their sides with one pectoral flipper and half of their tail flukes waving about in the air to scoop up massive amounts of burrowing ghost shrimp. These shrimp are about three inches in length and are overlooked as a food source by people since they don't carry enough fat to make them tasty, but the whales find them meaty compared to some of the other minute critters they ingest. While eating in this manner, they carve large impressions in the silt that resemble small bomb craters called "whale pits." Sometimes five or six will feed in this manner simultaneously, in a writhing, flipper-waving spectacle. The danger of stranding in this shallow water is real, but the dozen or so Sounders that have developed this feeding technique seem to have figured out the tidal flows and come in at just the right moment. The trick is not to stay too long.

In the town of Langley, Washington, there is a whale center and a park called Whale Bell Park. When the Sounders are spotted, locals are encouraged to ring the bell to alert everyone that the local celebrities have arrived. Each of the Sounders has been photographed and cataloged and the townsfolk think of them as personal friends. Some of the same whales have been returning since the early 1970s, surviving at least two major mortality events. The Sounders' deviation from the hereditary migration route seems sufficient proof of their ability to adapt and adjust. If this small pod of grays can adapt, then there is hope for others who may carve out little enclaves of survival while the ocean around them changes.

~

Many people are not aware that sewage from large cities flows directly into the ocean. There are also oil spills, coastal development, mercury pollution from the use of coal, and dead zones near river mouths from the flow of used fertilizers, but the worst offender remains the use of fossil fuels.

The United Nations' statement on climate change credits the ocean with absorbing up to 90 percent of the heat from rising global greenhouse emissions. Since the industrial revolution, it has mainly been humankind's burning of fossil fuels—coal, crude oil, and natural gas—that is raising the acidity of our oceans. Using fossil fuels increases the carbon dioxide ($CO2$) released into the atmosphere, and the ocean absorbs about 30 percent of the total output and stores it in plants such as seagrass. When carbon dioxide combines with seawater, it forms carbonic acid ($H2CO3$). The absorption of $CO2$ by seawater increases the concentration of hydrogen ions, which without getting very technical, directly translates to higher ocean acidity.

As seawater becomes more acidic, it loses the ability to hold calcium. Sea creatures, including coral, require calcium carbonate to grow their exoskeletons and protective shells. Marine critters can still grow their armor, but it will be brittle and take longer to develop. This mainly affects such creatures as clams, sea urchins, oysters, and corals, plus calcareous plankton, all of which are part of the food chain.

While the San Ignacio Lagoon has no coral, it does have plenty of shellfish, such as lobster and crabs. Even a place thought to have some of the purest air on Earth, such as the lagoon, is in fact slowly being choked by pollutants from the surrounding atmosphere. In just the last couple of years, marine science has coined the term "Trophic Cascade." It refers to minute ocean life that resides in the photic zone near the ocean's surface where photosynthesis takes place. Here is a brief explanation.

Whales feed at depth but return to the surface to poop, releasing vast plumes of fecal matter that is rich in iron (FE) and nitrogen (N), which nourishes plant plankton in the photic zone. This nutrient-rich fecal matter slowly sinks (cascades) to the bottom, where the continually diving and surfacing of whales mixes it and brings it back to the surface, thus allowing photosynthesis to take place for a longer period. This in turn helps to reduce acidity and CO2 emissions.

Some biologists have suggested that this continual mixing is the equivalent of that done by all the tides, waves, and currents, around the world. This surface plankton absorbs carbon (C) from the atmosphere and takes it deep to the bottom where it sits, harmless.

Before tens of millions of whales were slaughtered, uncountable tons of carbon were removed from the atmosphere by their simple daily routine of rising and diving. If humans were to cease hunting whales, this process would be increased greatly. Just by whales being whales, these animals directly help to slow climate change. I find this incredibly encouraging.

≈

According to NOAA, the National Park Service, and a study by Clemson University, as much as 90 percent of the trash currently in our oceans is plastic. Water bottles, fishing line, drinking straws, grocery lids, and shopping bags can easily travel from a local sewer to the ocean. It does not biodegrade. The longer plastic is exposed to the sun the more it separates into ever smaller pieces in a process known as photodegradation. Microplastic is categorized as any particle with a diameter less than five millimeters, and it is toxic to all living creatures that ingest it (all of humankind ingests it almost every day).

Microbeads, a category of microplastics commonly used in home hygiene products like facial scrubs, exfoliants, toothpaste, and hand soaps, are part of this problem. In 2015, then president Obama signed a law that went into effect in 2018 to phase out the use of solid plastic microbeads in rinse-off personal care products, but still today, some eight billion microbeads enter the environment every day. Meanwhile, synthetic clothing—especially rayon and polyester—sheds microfibers when washed.

Water treatment plants filter some of this but it is not enough. NOAA and the park service took samples from thirty-seven beaches in thirty-six national parks and resorts and found microplastics at every single site, with the highest concentrations being near the Great Lakes. Microfibers from synthetic clothing made up 97 percent of the plastics found.

Earth.org has reported on a necropsy performed on a beached whale in the Netherlands back in 2012. It was the first serious scientific investigation into the ingestion of plastics by whales. Forty-five pieces were found in the whale. Today, the ocean holds more plastic than ever before, and Earth.org estimates that each day whales probably ingest up to three million pieces of plastic. Some researchers analyze their feces to determine how much plastic is in any given area of the ocean.

There are mind-staggering heaps of trash and plastic in gyres, which are rotating systems of large ocean currents. Currently there are two immense ones in the Pacific Ocean. One is known as the Eastern Garbage Patch, and it extends from California to Hawaii. The other is the Western Garbage Patch, and its domain is from Hawaii to Japan. It is almost impossible to detect by satellite or even by boat when a visitor is in one of the patches because they are made up of incredibly tiny particles dispersed over vast areas. Much of it is smaller than a fingernail. The Ocean Cleanup Project estimates the Pacific Ocean gyre extends over 620,000 square miles (1.6 million square kilometers), and much of it has been in place for up to five decades. Since 1945 the numbers indicate both patches have grown exponentially—when combined, they are twice as large as the state of Texas. They also estimate that there are six pounds of plastic for each pound of plankton. This is what marine mammals are ingesting every day.

~

As maritime technology advances and modern cargo and cruise ships grow larger and faster, ships striking whales, especially grays, has become an ever-increasing possibility. The bridge (command center) of an average cargo vessel may be as high as ten stories. From there, it is impossible to see what is directly in front of the vessel, especially if it happens to be a whale that just surfaced. These ships travel at a sustained speed of anywhere from twelve to twenty-eight knots per hour, a knot being a nautical mile, and the equivalent of 1.85 kilometers per hour. Some of the more modern ships can motor as much as thirty-five knots per hour on the open sea. Such a vessel takes approximately two miles to stop while operating at full speed and more than a mile to turn. The gray whale swims at three to five knots per hour, not much faster than a human can walk. They do not echolocate to "see" what is around them, and the noise generated by the screws of a large vessel is more than enough to confuse a whale as to the direction it is coming from. The eastern coast of the rim of fire is a major shipping route along the West Coast of the United States, the same route followed annually by migrating gray whales. When big and fast meets slow and cumbersome, disaster is inevitable.

Gray whales navigate by underwater geography, so they migrate near the coast, both while going south and returning north. According to the National Oceanic and Atmospheric Administration, almost 50 percent of container ship traffic leaving or coming into the United States comes through the ports of Seattle, Portland, San Francisco, Los Angeles, Long Beach, and San Diego, all on the migration route of the gray whale. From 1988 to 2012, one hundred documented whale ship strikes were recorded. Not all of them were gray whales, but they did account for the most strikes. In 2019, five gray whales were necropsied in San Francisco Bay and all of them declared dead by ship strike.

Just days before the writing of this chapter, the Marine Mammal Hospital, the largest of its kind in the world, based in San Francisco, confirmed through their center that a whale had died from "blunt force trauma"—in other words, a ship strike. Three additional gray whales were found dead within a week. Necropsies were ordered for all of them.

There is no way to accurately record the number of whales killed by ship strikes. A large cargo vessel will not feel or hear anything when moving at flank speed on the open ocean. A whale might be struck and carried for miles on the massive wave cutters modern cargo ships have on their prows, or it might simply rupture and sink as seawater rushes in. A particularly vulnerable area is the Santa Barbara Channel of Southern California where the whales are funneled between the Channel Islands National Park and the mainland. In recent years the maximum speed limit through this channel has been reduced and shipping lanes altered to better accommodate the whales, but other than that, there is little that can be done. NOAA has made the following recommendations that are only common sense: 1) Learn the seasonal abundance of whales in the shipping lane and be aware of whale advisories. 2) Check with the navigation center of the US Coast Guard for animals known to be in the area. 3) Post dedicated, around-the-clock watches on the bow, with the single purpose of watching for whales. (This seems highly impractical, and perhaps impossible, especially at night.) 4) Reduce speeds in advisory zones, especially during the migration season, and 5) Reroute if possible. Still, as long as we occupy the same areas as the animals, there will continue to be conflicts.

~

In early February 2022, fishermen and pangeros in the lagoon reported sighting orcas. It was a first. The local English-language newspaper aimed at American expatriates, the *Gringo Gazette*, ran with the story and overnight the San Ignacio town and the lagoon were in a buzz.

For years I held the belief that being an apex predator, the orca would not enjoy making such easy kills as are available in the lagoon. I also believed the high saline content of the water would be a natural deterrent. Both theories have proven to be wrong.

I arrived in the lagoon only a few days after the incident, and while I spoke to several people who saw the orcas, I could not find anyone who witnessed them make a kill. However, a large chunk of gray whale blubber recovered by one of the pangeros may indicate that they did feed on the grays. It was an oddly shaped piece of flesh, about two feet in length and quite thick. It did not carry any recognizable bite markings. The orcas were spotted two days in a row and then apparently left, as there were no further reported sightings.

At the time of my last visit, everyone in Campo Cortez was holding their breath and praying it was a onetime event. The lagoon has always been the gray whale's sanctuary from orcas, and if they were to return with any regularity or bring other pods with them, thousands of years of gray whale habitat could be destroyed in a very short time. In addition, the local fish stock would be wiped out, ending both the whale-watching and fishing industries. It would be a devastating blow to the residents of the lagoon, not to mention all of central Baja.

≈

As modern cargo and cruise ships become larger and faster, the noise they generate increases exponentially. Sound travels three times faster underwater than it does on land, and that is an ever-increasing threat not just to cetaceans, but to all marine life. Whales have three to five times the brain cells dedicated to hearing than humans do, making them highly vulnerable to any underwater noise. A whale knows from miles away when a boat puts into the water or when an engine starts. Besides the noise of ships, there is military sonar, air guns used for seismic oil exploration, and offshore oil rigs. Mammal strandings are frequent after the deployment of naval sonar, which can disrupt an echolocating mammal such as a dolphin or orca from as far as 1,865 miles (3,000 kilometers) away. Such high noise levels can damage the inner organs and even cause vascular hemorrhaging of the brain. Further, a deep-diving mammal may panic during such noise deployment and surface too quickly, causing what divers call "the bends"—a buildup of nitrogen bubbles in the blood that can lead to death from an embolism.

Ocean creatures that do not echolocate depend on hearing just as humans do. Extreme ocean noise can disrupt schooling fish, interfere with finding a mate, or even with awareness of predators. One undocumented study has suggested that ocean noise causes cell changes in lobsters and has impaired the growth of shrimp. My personal observations have shown me how whales have been repelled by the sound of an engine, and researchers from Boston University have found that beaked whales seem to be the most susceptible to noise.

Christopher Clark of the K. Lisa Yang Center for Conservation Bioacoustics at the Cornell Ornithology Lab has observed that several decades ago, when there was far less ocean noise, whale communication usually only ceased during a storm, and resumed again once the weather passed. In more recent times, he has noticed that whales not only stop talking to one another when near oil exploration companies, but often evacuate the entire area.

Blue whales employ an extremely low-frequency voice that can be heard by another blue on the other side of the world. A humpback can hear one of its kind from several miles away, and North Atlantic right whales can communicate from up to ten miles away. Gray whales are among the softest speakers, and it's uncertain how far away they can communicate with others of their species. But the farther away the whales are from one another, the more their speech will be polluted by ocean noise. The National Oceanic and Atmospheric Association showed that from the 1960s through the early 2000s, there was a ten-decibel increase in underwater sound levels at low frequencies, the same frequency range that most whales communicate within.

~

Strange as it may sound, it is possible to love something too much, and in the case of the gray whale, it is an ever-increasing problem. From southern Alaska all the way south, the whales must run a gauntlet not only of marine predators, but also whale watchers who now take to the water by the tens of thousands. Almost every coastal town along their migration route has whale-watching tours, some of them taking multiple boats out multiple times a day. On any given day, that means thousands of additional engine noises are added to the regular shipping and ferry traffic. The Marine Mammal Protection Act is only as good as those who follow its regulations.

Phosphate deposits not far offshore, which grab the attention of mining companies, also pose a threat to the lagoon.

In 1994, Odyssey Marine Exploration, a deepwater search team, was founded in Tampa, Florida. It has surveyed and mapped close to twenty-five thousand square miles of the ocean floor, and in so doing, discovered tons of lost precious metals. They have also uncovered hundreds of shipwrecks dating back almost five thousand years. They found the *Black Swan*, rumored to contain four hundred thousand pounds of silver and gold, and the HMS *Victory*, Lord Nelson's flagship during the Battle of Trafalgar, among numerous others. In 2015 they decided to shift gears to become a search engine for the exploration and mining of undersea minerals. A mere 15.5 miles (25 kilometers) due west of San Ignacio Lagoon, in the Gulf of Ulloa, a vast phosphate reserve was discovered by the Mexican company Exploraciones Oceanics—and EO is partnering with Odyssey Marine Exploration on a mining operation called the "Don Diego Sea Mining Project."

Phosphate rock produces phosphorous, a mineral used in everything from pet and animal feed to electronic equipment to cosmetics. It is organic, cannot be artificially produced, is in everything we consume, and is essential to life as a part of the DNA cell wall. The massive phosphate deposit lies in forty-three to fifty-six feet (seventy to ninety meters) of ocean deep outside Mexico's territorial waters but within Mexico's Exclusive Economic Zone. The size of the deposit has recently been estimated to be as much as six hundred million tons. Since 2015, authorization has been denied by the Mexican government to begin the mining operation. However, Mexico has refused an initiative sent to its congress saying such an operation would put marine mammals at risk. Jorge Urban from the Autonomous University of Baja California Sur, who is the director of the Marine Mammal Research Program, has diplomatically argued that the mining of the phosphate would "modify" the migration route of the gray whale. Local fishermen have also joined forces with environmentalist Alejandro Olivera from the Center for Biological Diversity in Mexico City, who has petitioned the government to deny any future mining permits.

Odyssey Marine Exploration reports that its mining operation would have minimal impact on the local environment, arguing that no chemicals would be employed in the extraction and that all extracted sand—which it says would have no toxic effect on marine organisms—would be returned to the ocean floor after being dredged. It further states that the noise produced would be comparable to that of whale-watching boats and local fishing vessels. Its final statement is that the operation would have no impact on fishing because there is a traditionally low fish count in the area. Odyssey Marine Exploration has a majority ownership in Exploraciones Oceanics, and litigation continues while the project remains in limbo. It is ridiculous to state that such an enormous operation would have no effect on migrating sea life that will be in close proximity to it. The final decision as of this writing is yet to be known, but the constant threats to the pristine environment of the lagoon continue.

ELEVEN

The Great Mitsubishi Battle

No sooner had the Mitsubishi Corporation acquired 49 percent of the salt mining operations in Ojo de Liebre Lagoon than it began to set sights farther south on San Ignacio. A public relations campaign was launched in 1973 to convince the public that building the world's largest salt evaporation facility in the ancient calving and nursery lagoon would be both harmless and an economic boon to the local community. Along with its Mexican partner, Exportadora de Sal S.A. (ESSA), Mitsubishi's plan envisioned an industrial complex spread over an area of almost ninety miles. The goal would be the annual production of seven million tons of industrial salt, destined for the production of PVC piping in Japan along with chlorinated compounds. As explained by Professor Mark Spalding from the University of San Diego, who had been meticulously tracking the goings-on of Mitsubishi, "If you want chlorine to bleach paper, you want salt from Baja."

To meet Mitsubishi's goal, earthen dikes would be built over 116 square miles of evaporation ponds and a contingent of 17 diesel engines would pump 6,000 gallons per second out of the lagoon. Finally, a pier 1.25 miles long—intruding on traditional lobster and abalone grounds—was planned for loading the salt on container ships. According to Mitsubishi's projections, this would create a total of 210 jobs, half of which would go to local Mexican nationals while the rest would be trained and imported Mitsubishi workers from other locations.

When the salt works was originally installed in the northern lagoon, the whales instantly ceased going to that area and instead headed for Scammon's Lagoon. The proposed works for San Ignacio would be much larger in an area much smaller. The noise alone would prevent the whales from entering the lagoon.

Imagine an animal that has just traveled seven thousand miles, is physically exhausted, and greatly reduced in weight, reaching its destination, on what had been imprinted on its DNA for centuries, only to find it "under construction." Like any migrating animal, it would not know what to do. Their inner compass had brought them there, and there are no alternative places to go. Pregnant females would have to bear their young in open waters filled with predators, and there would most likely be widespread panic as more and more whales arrived to find their ancestral home destroyed.

In a *New York Times* article written in 1995, the Mitsubishi Corporation stated that for the past two decades it had been "operating in harmony with the whales," a reference to its already massive presence in Ojo de Liebre Lagoon. Not long afterward, the environmental watchdog CorpWatch claimed to have acquired a container of ESSA-produced salt whose consumer label specified it was "produced through a natural process completely compatible with the environment." The container had an image of a spy-hopping gray whale on its label. ESSA exports industrial salt to the United States, New Zealand, Canada, and several South American countries. Chlorine production emits an organic pollutant called dioxin, used in manufacturing PVC piping, that would be toxic to the whales. The effect the Mitsubishi plan would have on the whales would be a small indicator of the impact it would have globally.

Before 1973, Ojo de Liebre had been as pristine as San Ignacio. Now it had an infrastructure two times larger than San Francisco, and to many, it was a curse upon the land. Besides threatening the hereditary calving/nursery of gray whales, such an enterprise would also harm other aquatic wildlife such as the turtles and dolphins, plus the scallops, clams, abalone, and lobster, not to mention the terrestrial wildlife that roams free in the surrounding desert. Such an enterprise in San Ignacio would require miles of roads, housing, electrical, sanitation, and all the other infrastructural trappings of a large company town.

A full two years are required to produce usable salt from evaporation. Ocean water contains 3.5 percent salt. As the water is pumped out of the lagoon over a period of eighteen months, it moves through a series of ponds where the salt is concentrated through wind and solar-aided evaporation. This raises the salt content to roughly 27 percent. That water is then stored in shallow ponds for the six months it takes for the salt to crystallize. From there, barges take the salt to ESSA facilities on Cedros Island for shipment on cargo freighters out into the world. ESSA had deemed San Ignacio Lagoon ideal for this process due to little rainfall and vegetation, with a high rate of evaporation.

When news of this planned salt works became public, environmental groups from around the world went to work. Public pressure forced Mexico's environmental agency to reject the original environmental studies and create a bipartisan panel of international observers to reevaluate the entire project. In 1995, the National Resources Defense Council stepped in to lead an international coalition in opposition to the proposed plan.

In 1997, Mitsubishi and ESSA issued a joint statement declaring their proposed plan had been redesigned to address international concerns. During this time, rumors spread that Mitsubishi planned to close its operation at Ojo de Liebre and transfer all works to the new and more expansive site at San Ignacio, expanding the destruction exponentially.

During this time, three hundred violations of twenty-two different laws were filed against Mitsubishi, prompting the environmental coalition to act. The cities of Los Angeles, San Francisco, Berkeley, Sacramento, and Poway all issued proclamations condemning the new salt works while more than fifty Mexican environmental groups filed lawsuits against Mitsubishi. Then the public got onboard. The Natural Resources Defense Council generated more than one million letters to the Mexican government and Mitsubishi in a little more than two weeks. The coalition petitioned UNESCO, warning that one of their World Heritage Sites was under attack and enlisting nine Nobel laureate scientists to condemn the new salt works. They enlisted crack Mexican lawyers and diplomats to advise and manage the campaign and procured resolutions condemning the operation from state agencies all along the coast, including the California Coastal Commission.

The famed Mexican poet, diplomat, and environmental leader Homero Aridjis rallied his El Grupo de Cien (the One Hundred) composed of artists, politicians, actors, and the more enlightened of Mexico's literati. For several days, the hotels and restaurants of tiny San Ignacio filled with some of the most heavyweight environmentalists on Earth. Campo Cortez was so full that old emergency canvas tents were set up to accommodate the overflow. On any given day of that week, you might have shared a street taco with a politician, movie star, or world-renowned artist. The One Hundred were joined by such illustrious names as Octavio Paz,

Alan Ginsberg, Margaret Atwood, both Sir James and Sir Edward Goldsmith, Peter Matthiessen, and Tom Hayden. Representatives from thirty-five diverse environmental organizations, such as Greenpeace, the Humane Society International, Cetacean Society International, and the Animal Welfare Institute, were in town. Shuttles ran day and night between San Ignacio town and the lagoon.

The infamous founder of Sea Shepherd, Paul Watson, of the Makah whale hunt fame, was anchored offshore outside the lagoon in international waters since his rather violent track record of anti-whaling tactics had made him persona non grata in Mexico. He kept his distance and used a false name on the radio to keep track of events while not actively participating.

The Natural Resources Defense Council, led by attorney Joel Reynolds, joined the fight, along with Mexico's ambassador to the United Nations, Manuel Tello Macías. American celebrities and activists such as Glenn Close; Robert Kennedy Jr.; Pierce Brosnan and his wife, Keely; and Jean-Michel Cousteau, son of the famous oceanographer, descended on Campo Cortez along with noted scientists such as cetacean researcher Dr. Roger Payne and conservationist photographer Robert Ketchum, who spent alternating weeks giving interviews and appearing at press gatherings. With such a public surge, the opposition spread faster than an incoming tide. Money fund managers were recruited to state that they did not consider Mitsubishi to be a good environmental investment, and a public economic boycott of the company in all its various forms was initiated. Mitsubishi, whose environmental track record was abysmal, saw the writing on the wall. The world had said no to its salt works.

On March 2, 2000, then outgoing Mexican president Ernesto Zedillo held a press conference in Mexico City to tell the world that Mexico was abandoning its plan for a salt works in San Ignacio Lagoon.

On December 1, 2000, Vicente Fox was elected the new president of Mexico and wanted to bring his granddaughter to the lagoon to see the whales. With word of his impending visit, the dirt, rough-graded road that connected San Ignacio town to the lagoon was suddenly paved for almost ten miles just outside of town in a matter of days. It was said that after his granddaughter petted a whale, he personally committed to the preservation of the lagoon.

≈

In 2005 the Laguna San Ignacio Conservation Alliance was born to secure lands surrounding the lagoon. The NRDC and its partners have acquired conservation easements of almost 150,000 acres of land bordering the eastern side of the lagoon while Pronatura Noroeste secured a conservation concession of nearly 200,000 acres on the west side from the Mexican government. Another environmental organization called WILDCOAST has a concession to guard the mangrove forests and undeveloped beaches of the area. As of 2020, more than $2 million has been put into a trust to provide annual payments for the continued monitoring of the area along with developing local community projects.

According to the NRDC's Joel Reynolds, who is the western director and senior attorney for Marine Mammals, Oceans Division, Nature Program, "This grassroots coalition, working together, has secured the largest environmental victory ever."

TWELVE

One Final Whale Story

This may be one of the most unusual whale stories you will ever read. When I am not working as a naturalist, I have for years enjoyed visiting remote tribal cultures to collect stories. Years ago, I spent several weeks on an expedition riding through the Sahara Desert of landlocked Mali with Tuareg nomads, members of the Berber tribe who have a reputation as fierce warriors. When we reached our destination of Timbuktu, I wished to offer a proper thank-you to my two intrepid guides, but money did not seem personal enough. Just before I went on that trip to Africa, I had given a lecture about gray whales and still had the program on a portable hard drive. I thought it would be great to show them a creature they had never seen, so I connected my hard drive to the nine-inch, black-and-white television in my room and invited those two hardened desert warriors to join me. They sat cross-legged on my bed in their regal blue robes and turbans with long curved daggers stuck into their waist sashes, wondering what was about to happen. They were men of the deep desert and not used to being in hotel rooms, let alone in the presence of modern amenities such as a television.

As I showed the images of gray whales breaching and spy hopping, they were at first confused and asked if they were fish. Realizing they had no reference for size, I knew they had seen hippos in the Niger River and told them these oceanic beasts were called whales and were many times larger than a hippo. They stared at me wondering if I was telling a joke, because the Tuaregs love a good joke. After a few moments they looked at each other and burst out laughing, then began to pound me on the back yelling "big fish!" over and over while pointing at the screen and laughing heartily. Apparently, it was the most ridiculous thing they had ever heard. When I finally convinced them that the images were actual creatures, they both approached the television from only inches away to take a better look, still baffled. The elder of the two ran his hand over the screen in hopes of an actual touch. We bid our goodbyes, and they walked off into the night on the sand-covered street, laughing loudly about the "big fish." I can imagine them trying to describe "big fish" to landlocked desert nomads who do not eat fish and have probably never actually seen one. I am sure that story grew with each telling, especially at my expense.

The next day, my final one in Timbuktu, I stopped at the central bazaar and found a small piece of gray whale baleen. When I picked it up, the lady vendor took it back from me and explained how to use it as a broom. Today it resides in my office, a most unusual souvenir from the Sahara Desert. Whenever I look at it, I wonder about the journey it made from San Ignacio Lagoon.

Epilogue

Long ago a scientist friend told me the ocean always heals itself. I believed that for many years, but now I am no longer sure.

I have witnessed climate change firsthand as it alters migration patterns and feeding habits. I have watched our sacred ocean become a dumping ground by thoughtless humans.

When I take people out on a boat, I give a safety lecture followed by a request not to throw anything overboard. Even though I use photos of animals trapped by plastic to illustrate the threat, there is always someone who casually tosses their potato chip bag or soda bottle into the water. When I see this happen, we stop the boat and turn around to pick up the offending trash. I never say a word because they are paying customers and I have no right to lecture anyone, but most of them finally understand when we stop their whale tour to retrieve their trash.

Where I currently live on the central coast of California, there was a land-based whaling operation from 1916 to 1927. During that time there is no way to know how many whales were taken. I am proud that, today, the same operation is an intellectual center for marine biology, powered by one of the nation's finest aquariums, which maintains enormous research vessels that share ocean knowledge in real time with every aquarium and university on the rim of fire. It is also a hugely popular whale-watching harbor. Moss Landing, California, is as fine an example of a onetime hunting ground becoming a conservation center as is San Ignacio Lagoon.

After almost three decades on the water, I have come to believe there may be hope for the enlightenment of mankind. Mass media, for all its shortcomings, has played a key role in assimilating and dispensing information that I believe has directly contributed to the public education regarding whales. The evidence is overwhelming that these creatures feel pain and sorrow, joy and elation. They love just as people do, and they exhibit empathy that rivals ours. For all we know, they may think of us as the inferior ones. Whales traveled this Earth for countless centuries before humans came on the scene, and they have survived near-extinction events over and over. They were cruelly hunted to the brink and still came back to forgive us and to offer their friendship. There is no other creature on Earth as evolved as that.

San Ignacio Lagoon is a holy place, a cathedral in the wilderness the equal of any great church constructed by humankind. A Lakota Sioux chief once said that most men go into a building to talk to God, while some of us go into the wilderness and God speaks to us. God has always spoken to me in the lagoon. While the lagoon's creatures are giant in size and strength, they are still at the mercy of humans. Like it or not, assuming the responsibility or not, we are the stewards of the lagoon, and we have the power to destroy it or preserve it.

It is my fervent hope and prayer that my friend was right in saying the ocean always heals itself. But each year as I leave, I cannot help but feel it may be the final time, that it is so ethereally fragile that it cannot possibly withstand the constant onslaught of modern society. I pray that I am wrong.

For all of the above, I am forever indebted to the people who call the lagoon home, particularly those at Campo Cortez. They are knights who stand watch and tell the animals nothing will harm them while they are on duty. They have been my teachers, my guides, and my friends. Without them, this book could not have been written.

Acknowledgments

I am forever in the debt of the gray whales who made this book possible. My eternal love and respect go out to my wife, Irene, who has been at my side on land, sea, and air, for more than five decades. I can never be grateful enough for my Baja family in Campo Cortez. Maldo, Catalina, Elsa, Paco, Roberto, Cuco, Valentin, CeCe, Gadriel, and Jennifer graciously opened their lives and their souls to the world. A special thank-you goes to my publisher Keith Wallman, and his right hand Jane Glaser, whose enthusiasm for this book was an inspiration. To all the crew at Diversion Books whom I have not met, thank you for your part in this collaboration. Finally, my gratitude is unbounded for my agent, Lisa Hagan, whose vision started this whole enchilada rolling.

Index

About the Author

James Michael Dorsey is an award-winning author, explorer, marine naturalist, and lecturer, who has traveled extensively through fifty-six countries. He has published three collections of his travel narratives, and written for *Colliers*, the *Christian Science Monitor*, *BBC Travel*, *BBC Wildlife*, *Lonely Planet*, United Airlines in-flight magazine *Hemispheres*, *Los Angeles Times*, *Chicago Sun-Times*, *California Literary Review*, and numerous African magazines. His writings about the natural world strive to educate and entertain those who have not had the privileged access he has been gifted, as his pursuit of whales has taken him to three countries and five states. His stories about his remote travels are an attempt to give some small voice to those with no written language and to let the world know these vanishing cultures exist. He is a longtime member of the American Cetacean Society, a fellow of the Explorers Club, and member emeritus of the Los Angeles Adventurers' Club. Near his California home, he volunteers with Marine Mammal Rescue Service. He is also grateful to be represented by Lisa Hagan Literary. He has been married to his costar in this book, Irene, for fifty-two years.

www.jamesdorsey.com